Artificial Intelligence in Pharmaceutical Sciences

This cutting-edge reference book discusses the intervention of artificial intelligence in the fields of drug development, modified drug delivery systems, pharmaceutical technology, and medical devices development. This comprehensive book includes an overview of artificial intelligence in pharmaceutical sciences and applications in the drug discovery and development process. It discusses the role of machine learning in the automated detection and sorting of pharmaceutical formulations. It covers nanosafety and the role of artificial intelligence in predicting potential adverse biological effects.

FEATURES

- Includes lucid, step-by-step instructions to apply artificial intelligence and machine learning in pharmaceutical sciences
- Explores the application of artificial intelligence in nanosafety and prediction of potential hazards
- Covers application of artificial intelligence in drug discovery and drug development
- Reviews the role of artificial intelligence in assessment of pharmaceutical formulations
- Provides artificial intelligence solutions for experts in the pharmaceutical and medical devices industries

This book is meant for academicians, students, and industry experts in pharmaceutical sciences, medicine, and pharmacology.

Artificial Intelligence in Pharmaceutical Sciences

Edited by
Mullaicharam Bhupathyraaj
College of Pharmacy, National University of Science and Technology,
Muscat, Sultanate of Oman

K. Reeta Vijaya Rani
Surya School of Pharmacy, Vikravandi, Tamil Nadu, India

Musthafa Mohamed Essa
Sultan Qaboos University, Sultanate of Oman

CRC Press
Taylor & Francis Group
Boca Raton London New York

CRC Press is an imprint of the
Taylor & Francis Group, an **informa** business

First edition published 2024
by CRC Press
2385 NW Executive Center Drive, Suite 320, Boca Raton FL 33431

and by CRC Press
4 Park Square, Milton Park, Abingdon, Oxon, OX14 4RN

CRC Press is an imprint of Taylor & Francis Group, LLC

ISBN: 978-1-032-36386-8 (hbk)
ISBN: 978-1-032-38215-9 (pbk)
ISBN: 978-1-003-34398-1 (ebk)

DOI: 10.1201/9781003343981

Typeset in Times
by KnowledgeWorks Global Ltd.

Contents

Preface

Pharmacy is an active scientific field related to human and animal healthcare that comprises the original and innovative discovery, formulation, standardization, manufacturing, monitoring, and regulation of drugs. For the past few decades, the pharmacy field required unique expertise to assist and support the educational facets and pharmaceutical sector/industries.

Artificial Intelligence (AI) has offered a crucial part in significantly developing and nurturing the field of pharmacy throughout the world. The chapters in this book will deal with the omnipotent and omnipresent inevitability of AI, an overview of AI-driven pharmaceutical functionality, AI in drug delivery systems, the role of AI in drug discovery and development, AI applications in scrutinizing health status, and the benefits of robotics in surgery, analysis of pharmaceutical formulations using AI, role of AI in pharmaceutical drugs, gene therapies, and medical device regulation, ethical aspects of applications of AI in medicine and pharmacy, role of machine learning in automated detection and sorting of pharmaceutical formulations, AI in community pharmacy, and AI is all about creating a global win-win situation in the pharmaceutical world to enhance healthcare.

This book will be very useful in an academic setting (students, faculty, staff, researchers, and scientists), pharmaceutical industry, and regulatory affairs. Thus, this book will clearly and rationally convey and establish the impact of AI in the pharmacy world.

Mullaicharam Bhupathyraaj
K. Reeta Vijaya Rani
Musthafa Mohamed Essa

Editors

Mullaicharam Bhupathyraaj, PhD, is Professor of Pharmaceutics at the College of Pharmacy, National University of Science and Technology, Muscat, Sultanate of Oman. She has more than two decades of teaching and research experience and has received awards at the national and international levels. Professor Bhupathyraaj received many research grants from HRD, AICTE, RG, and MoHERI (NRG, TRC) Oman and has eight international patents to her credit. Her research interests include biomedical, nanopharmaceuticals, herbal formulation development, and applications of artificial intelligence (AI). She has publications and patents in the AI field in medical imaging. Professor Bhupathyraaj has published more than 90 research articles in peer-reviewed journals and several book chapters, serving as an editorial board member and a reviewer in various peer-reviewed journals and presenting many papers at various conferences in India, the UK, Italy, the United States, Oman, and UAE. She has authored many textbooks (USA and Germany). She is an external examiner for PhD theses of various universities and a reviewer of peer-reviewed journals. Professor Bhupathyraaj is a Registered Pharmacist in India and a life member of professional bodies of IPA, APTI, IPGA, and ISTE. She has organized many seminars and conferences. She is the external reviewer of the Oman Academic Accreditation Authority, Sultanate of Oman.

K. Reeta Vijaya Rani, PhD, is Professor in the Pharmaceutics Department at Surya School of Pharmacy, Vikravandi, Tamil Nadu, India, and a member of the Surya Group of Institutions. She has 26 years of experience in pharmaceutical research and education, and she has authored and presented papers at both the national and international levels. Professor Rani has extensive academic writing knowledge and experience. She has successfully completed over 58 projects, including papers, posters, and national and international conference presentations. She has written chapters and textbooks on pharmaceutical science for national and international publications and textbooks. Professor Rani has mentored over 200 pharmacy students with their research papers on novel topics at the postgraduate level. Professor Rani is an external examiner for PhD theses of various universities and a reviewer of peer-reviewed journals. Her key research interests are biodiversity and novel therapeutic techniques. When it comes to institutional quality review, she is a wealth of information (ISO). Professor Rani is the owner of two patents for innovative pharmaceutical formulations.

Musthafa Mohamed Essa, PhD, is Professor (former) at Sultan Qaboos University, Oman. He is Editor-in-Chief for the *International Journal of Nutrition, Pharmacology, and Neurological Diseases*. He is also Associate Editor for *BMC Complementary and Alternative Medicine* and Editor of *Frontiers in Neuroscience* and *Frontiers in Biosciences*. Professor Essa is an expert in the field of nutritional neuroscience/neuropharmacology and has published more than 200 papers, 64 book chapters, and 14 books. He is the recipient of several awards, including the best book in the world, the National Research Award, Outstanding Neuroscientist, and Distinguished Researcher. Professor Essa has over 15 years of research and teaching experience in the field of nutritional neuroscience, and under his guidance, 5 PhD and 10 master's students completed their research. He is Editor of a unique and first-of-its-kind book series titled Nutritional Neurosciences, published by Springer Nature.

Contributors

Ronaldo Anuf A.
Department of Biotechnology
Kamaraj College of Engineering and
    Technology
Vellakulam, Tamil Nadu, India

Hanan Fahad Alharbi
Department of Maternity and Child Health
    Nursing College of Nursing
Princess Nourah bint Abdul Rahman
    University
Riyadh, Saudi Arabia

Carolin Lincy B.J.
Department of Pharmaceutical Technology
College of Engineering
Anna University
Chennai, Tamil Nadu, India

Mullaicharam Bhupathyraaj
College of Pharmacy
National University of Science and Technology
Muscat, Oman

Leena Chacko
Meso Scale Diagnostics LLC
Rockville, Maryland, USA

Muralikrishnan Dhanasekaran
Department of Drug Discovery and
    Development
Harrison College of Pharmacy
Auburn University
Auburn, Alabama, USA

Manisha Ganesh
Department of Pharmacy Practice
Nandha College of Pharmacy
Vailkaalmedu, Tamil Nadu, India

Prabhu Subbanna Gounder
Department of Computer Science and
    Engineering
Nandha Engineering College
Erode, Tamil Nadu, India

Ashok Kumar Janakiraman
Faculty of Pharmaceutical Sciences
UCSI University
Kuala Lumpur, Malaysia

Kushagra Khanna
Faculty of Pharmaceutical Sciences
UCSI University
Kuala Lumpur, Malaysia

Kiruba Mohandoss
Sri Ramachandra Institute of
    Higher Education and Research
Chennai, Tamil Nadu, India

Sriram Nagarajan
College of Pharmacy
Holy Mary Institute of Technology
    and Science
Hyderabad, India

Kirubakaran Narayanan
Department of Pharmaceutics
SRM College of Pharmacy
Kattankulathur, Tamil Nadu, India

Sobana Ponnusamy
Department of Computer Science and
    Engineering
Nandha Engineering College
Erode, Tamil Nadu, India

Yoga Senbagapandian Rajamani
College of Computer, Mathematics, and
    Natural Sciences
University of Maryland
College Park, Maryland, USA

Vijaya Rajendran
Department of Pharmaceutical
    Technology
College of Engineering
Anna University
Chennai, Tamil Nadu, India

Subha Sri Ramakrishnan
Department of Pharmaceutical Technology
College of Engineering
Anna University
Chennai, Tamil Nadu, India

Akila Ramanathan
College of Pharmacy
Sri Ramakrishna Institute of Paramedical
    Sciences
Coimbatore, Tamil Nadu, India

Gothanda RamanG
Department of Pharmaceutical Technology
College of Engineering
Anna University
Chennai, Tamil Nadu, India

K. Reeta Vijaya Rani
Surya School of Pharmacy
Vikravandi, Tamil Nadu, India

S Anbazhagan
Surya School of Pharmacy
Vikravandi, Tamil Nadu, India

Ramkanth Sundarapandian
Department of Pharmaceutics
Karpagam College of Pharmacy
Coimbatore, Tamil Nadu, India

Hemalatha Selvaraj
Department of Pharmacy Practice
Nandha College of Pharmacy,
Vailkaalmedu, Tamil Nadu, India

Umashankar Marakanam Srinivasan
Department of Pharmaceutics
SRM College of Pharmacy
Kattankulathur, Tamil Nadu, India

1 The Omnipotent and Omnipresent Inevitability of Artificial Intelligence (AI)

AI Is All about Creating a Win-Win Situation

Kiruba Mohandoss
Sri Ramachandra Institute of Higher Education and
Research, Chennai, Tamil Nadu, India

1.1 INTRODUCTION TO ARTIFICIAL INTELLIGENCE (AI) RESEARCH

AI research goals involve a broader perspective factor, including theoretical thinking, knowledge representation, planning, learning, natural language processing (NLP), and understanding the core concepts; thus, the overarching objective is to move and manage objects in order to address issues unilaterally and synchronously. AI scientists and researchers leverage mathematical version search, logical neural systems, statistics, and probability-based economics to dive deep into AI computer science, psychology, linguistics, philosophy, and continuum. In the nascent stages, thought leaders and early adopters morally justified creating philosophical debates centered around inventing entities with AI. The bigger picture was to harness the power of AI to optimize the functionality of diverse critical domains (1). But the concept of AI-powered entities has been around from times immemorial in the form of fairy tales, mythology, fantasy, and philosophy. It is now that the human race, supposedly the most intelligent species, has started reimagining the rudimentary concept of intelligence and its portability. On the flip side, some theories suggest that AI is dangerous to humans if used without help. And few others perceive AI as the opposite of the previous technological revolution, posing a tremendous unemployment risk. From an elementary standpoint, AI is a way of thinking linked to the mind of a person who has a computer or a robot or any digital product. AI is how the human brain thinks, learns, defines, and functions as it tries to solve problems. The bottom line of AI is all about digitally solving problems related to human understanding, such as thinking, comprehending, learning, and problem-solving (2). AI research synergizes science, planning, education, and NLP to infuse thinking ability into objects. Research methodology mostly includes combining statistical tools, intelligent calculations, and mathematical research to balance traditional thematic AI coding and AI research (3).

1.2 HISTORY OF AI

It is imperative to understand scientific paradigm changes and the history of AI to rake in the benefits through the outcomes. It helps determine the criticality of AI as a performance-enhancing tool that does not fully replace humans both on the job market and, most importantly, as final decision-makers. AI has evolved, during the last five decades, starting with a very classical approach grounded in mathematics and psychology. It was then followed by an era in which almost everything was perceived to be possible for a computer to solve. Intelligence has many facets, including problem-solving, learning, recognizing and classifying patterns, building analogies, surviving by adaptation, language understanding, creativity, and many others. Despite AI being around for millennia, it was only in the 1950s that its true potential was probed. A generation of scientists, physicists, and intellectuals was always fascinated by the concept of AI; it was Alan Turing, a British polymath, who was the first to propose that people address issues and make decisions using available information and reasoning. The lack of computers was the major stumbling block to further exploration of the concept. Researchers needed to adapt fundamentally before expanding any further. The then contemporary machines could execute orders but not store them, and until 1974, financing was also a problem. Toward the end of 1974, computers had become extremely popular and were considerably quicker, affordable, and capable of storing more data. The advent of computers with superior capabilities laid strong foundations for research work on AI, and the rest, as they say, is history!

In the following years, it was comprehensively proved that it was not too difficult to build systems and algorithms incorporating some intelligence (although far from encompassing all the possible facets). As Hutter stated in 2005, "Most, if not all known facets of intelligence can be formulated as goal driven or, more generally, as maximizing some utility function." A more pragmatic attitude led AI researchers to develop knowledge-based systems in which transparency and explainability were mandatory for the sake of real applications. This resulted in a modern trend based on age-old essentials in which adaptation, cooperation, learning, and autonomy become important features of more sophisticated intelligent systems. It was not very long ago that the traditional step-by-step AI systems development experienced disruptive interventions. This gave rise to new warnings about the potential dangers of possible misuse of AI algorithms and systems. This disruption was triggered by "big data," "Internet of Things" (IoT), and new algorithms like those related to the "deep learning" concept. This, in turn, led to the development of striking applications with huge economic and social outcomes (4).

Reflecting upon such outcomes is no longer a kind of an unnecessary distraction, "like worrying about the overpopulation of Mars," in the words of Andrew Ng, quoted in Ref. (5). The algorithms, usually called deep learning, mostly rely on the artificial neural networks (connectionist) paradigm. Connectionist-based methods approach has the big advantage of avoiding the knowledge acquisition bottleneck since the proposed model is directly built from observations with very little human intervention. The flip side is manifested in the form of a kind of black box. There are, however, a host of diverse approaches that AI researchers have followed in the past, which are responsible for relevant AI-based systems applications. In Ref. (6), the author identifies five AI and machine learning (ML) "tribes" that currently exist: the symbolists, connectionists, evolutionary, Bayesians, and "analogizers." Despite very different paradigms and different schools of thought at play, the objective is always the same: To develop machine intelligence. AI has been repeatedly overhyped in the past, even by some of the founders.

Consequently, the so-called AI winters(s) hit the field, decreasing the potential outcomes of AI realizations. Nevertheless, well-known researchers like R. Brooks, a critic of Good Old Fashion AI (GOFAI), have opposed the idea that AI failed. Brooks rightly predicted that AI would be around us daily (7). Hence, it is indeed imperative to recognize the outcomes AI has already achieved. Collectively, researchers should also reflect upon its potentiality and future social outreach. Nevertheless, intensive overselling that raises huge expectations of AI-based systems (without discussing their inherent dangers) must be avoided (8).

1.3 BASIC CONCEPTS OF AI

The evolution of AI as a change agent has been rapid all across the business landscape. The last few years have witnessed emerging AI concepts being rolled out at a frantic pace. Autonomous vehicles, big data, and medical research are some of the most awe-inspiring natural applications that have emerged from the advancement of AI. In fact, AI has been embraced as a process-optimizing framework in a wider and deeper way than expected. Hence, it is important to comprehend the basic concepts of AI to gauge its influence across diverse domains. From a very basic viewpoint, AI defines the ability of a personal computer (PC) or a machine to make and manage certain activities or decisions. AI architects aim to mimic human interpretations like creativity, decision-making, calculation, automation, etc. Virtual assistants, chatbots, and similar tools within different structures at different levels are classic examples of AI at play (9, 10).

1.3.1 Preliminary Features of AI

These include the following essential information:

i. *Categorization:* AI needs a lot of information about the problem it has to solve. Creating an effective AI framework requires creating a category or criteria for a specific field. These gauges are used by machines to identify issues and address them. Whether in healthcare, information technology (IT), the pharmaceutical industry, banking, or any other domain, the user needs to define the measurement features that allow the problem to be broken down into smaller parts (11).

ii. *Classification:* User issues or specific business needs are divided into different categories. The next step involves a classification for all classes that leads the customer to a critical end. For example, when training the AI framework for playing the most popular television game, "Who Wants to be a Millionaire," the administrator must feed the questions and then categorize them by time, subject, location, degree of difficulty, or any other relevant criteria. Once the type of problem is identified, the user must classify the cause of the problem: authentication, Dynamic Host Configuration Protocol (DHCP) binding, or other wireless devices (12).

iii. *Machine Learning:* ML is a part of AI and computer science that centers on the use of data and algorithms to model in what manner humans learn, gradually increasing accuracy. The world-renowned technology giant International Business Machine (IBM) has a long history in ML. A customized ML and research game was played by self-proclaimed monitoring expert Robert Nile in 1962, who played against an IBM 7094 computer and lost. This achievement seems incomparable to what can be accomplished today. Huge advancements in ML from a storage and computing perspective are helping innovators discover newer products like Netflix, recommendation systems, or self-driving cars (13). There are several ML techniques, and neural network–powered ML has become a major method involved in AI. It is further divided into supervised learning and remedial learning, where ML facilitates learning from experience without having to program the task. The process starts with quality data entry, then training the machine by making various models using various algorithms. The algorithm selection depends on the type of work to be automated. ML algorithms generally fall into three categories: supervised learning, unsupervised learning, and reinforcement learning (14).

iv. *Deep Learning:* Deep learning is a function of AI that mimics the activity of the human brain to create patterns for information processing and decision-making. It is a subset of ML in the field of AI that can control learning from unstructured or networked data. This is also referred to as deep neural learning or deep neural network. This allows users to process and predict data using neural networks. This brain-like network is connected to the human brain through a network-based structure. It is a ML function that works in a nonlinear decision-making process. In-depth learning takes place when deciding on unstructured unattended data. Object recognition is part of what is done by learning and translating the voice. Hierarchical neural networks are used to analyze data as part of ML in in-depth analysis. Neural codes are interconnected in hierarchical neural networks. The classification structure of advanced learning differs from other traditional thematic programs of linear machines similar to the human brain. It processes data on a set of layers, and each level provides more information for the next level, using in-depth training to detect deception or fraud. This system uses different signals, such as IP address, credit rating, retailer or sender, etc., in the first level of the neural network (15).

v. *Reinforcement Learning:* Learning empowerment is a part of AI where a machine learns something the way humans learn it. Just like students, machines can learn from their mistakes through test time and error. This means the algorithm learns behavior based on the current situation and will judge the next action by maximizing future income. A famous example of this is Google's Alpha Go computer program reinforcement learning, which defeated the world champion in the 2017 Go Games.

vi. *Robotics:* Robotics is a branch that ensures machines see and work like humans. Now, robots can act like humans in certain situations, but they cannot think exactly like humans. This is where AI comes in! Al helps robots to work smarter in certain situations. These robots can solve problems in confined spaces or even learn in a controlled environment. The social interaction robot Kismet developed at "MIT's Artificial Intelligence Lab" is a perfect example, and the Robnet, built by NASA to work with astronauts in space, is another classical example.

vii. *Natural Language Processing:* Naturally, people can talk to each other using words, and now machines can do it too. This is called natural language processing, where the machine understands and decodes language and dialect while speaking. There are many components in the NLP language, such as speech recognition, natural language structure, natural language translation, etc. NLP is leveraged to drive many customer support applications like chatbots. These chatbots use ML and NLP to communicate with users in a text format and solve a variety of problems. It is a technique that facilitates customer support without having to talk to the customer directly. Examples of popular NLP applications are Alexa from Amazon and Siri from Apple.

viii. *Recommender Systems:* Imagine the scenario where the device offers recommendations for movies and series based on specific preferences. This is triggered by the referral system, which advises on how to choose the next key from the many options available online. Curator systems can use content-based recommendations or even filter interactions. Content-based recommendations are achieved by analyzing the content of the overall show. For example, one can make a recommendation based on the show's description and one's basic profile.

ix. *Computer Vision:* The Internet is full of pictures or photographs. In this age of selfies, millions of pictures are uploaded and viewed on the Internet every day. This technique enables the computer to see and understand any picture. Computer vision uses AI to gather information from images. This information can identify the objects in the image (like group images with image content) and analyze the surrounding images for automated vehicles such as Mars landings and the automation used by Opportunity Rover.

x. *Collaborative Refinement:* Most have a group filtering experience when they select a movie on Netflix or buy something from Amazon and get recommendations. In addition to the recommendations, collaborative filtering is used to detect large datasets and manipulate AI. This is where all the classification and analysis of data is transformed into critical knowledge or activity. Collaborative filtering is a way to answer with a high degree of confidence, whether used in game shows or used by an expert or by the manager of an organization.

xi. *Internet of Things:* IoT is a holistic ecosystem of AI that leverages previous experience and learns to simulate human work without manual intervention. All IoT devices, including wireless sensors, software, actuators, computer devices, and more, produce a lot of data that must be collected and excavated. IoT is used to collate and analyze big data to formulate AI algorithms that can be used across various instruments (16).

1.4 WORKING PROCESS OF AI

Most tasks in many contemporary organizations are not automated. Intelligence-powered smart functionality is the need of the hour. AI in the enterprise domain can be defined as the ability of organizational intelligence to acquire knowledge and apply it to produce superior results. AI leverages organizational procedures, systems, and history to automate and simplify processes. This is stated in a comprehensive report published by the National Council for Science and Technology (NSTC): "Some have defined AI as a computer system that often behaves in a way that requires intelligence. Others have defined AI as a system that can logically solve complex problems or take appropriate action to achieve its goals." At the basic level, AI programming focuses on three learning skills: learning, reasoning, and self-correction (17).

i. The learning aspect of AI programming focuses on data acquisition and creating rules for converting data into action. The rules, called algorithms, provide a computational method with step-by-step instructions on how to perform a specific task.
ii. Reasoning relates to the AI capability of selecting the most appropriate algorithm, to use in a specific context in a set of algorithms.
iii. Self-correction focuses on AI's ability to improve and enhance results over time, until the desired goal is achieved.

Amidst the rapid growth of AI-related advertising, businesses are trying to promote how their products and services leverage the power of AI. Organizations must fully comprehend the nuances of deploying AI as part of their functionality. This mandates thorough knowledge of AI writing and training procedures. AI writing and training require a specific hardware and software foundation. Typically, AI systems work by receiving large amounts of labeled training data. Subsequently, they analyze data for relationships and patterns. These models are used to make predictions about future status. Tools like image recognition tools can learn to identify and describe objects in an image by examining millions of patterns. ML algorithms are not synonymous with any specific programming language. There are some popular options, like Python and Java, offering cutting-edge exposure to aspiring AI professionals (18).

1.5 ORIGINS OF AI

The modern field of AI dates back to 1956, when the term AI was used in a proposal for an academic conference at Dartmouth College. It was built around the assumption that the human brain can be controlled by a machine to produce smart results in a sustained format. Yet, it can be assumed that the very basic concept of AI in its most primitive form was formulated thousands of years ago. For example, many ancient cultures created rational and emotional automata like humans; this was closely linked to deep thinking. During the first millennium BC, philosophers around the world developed methods of formal thinking. These were efforts made by collaborators, including theologians, mathematicians, engineers, economists, psychologists, and neuroscientists, over the next 2,000 years. This phenomenon laid the foundation for what is today known as AI (19).

1.6 TYPES OF AI

As per David Peterson, one of the most eminent thinkers on the subject of AI, modern AI has evolved from rudimentary AI systems into frameworks capable of simple classification and pattern recognition that can predict actions using historical data. This phenomenon is possible because, within the contemporary ambit of deep learning, AI learns from big data or information. AI has been rapidly evolving through the twenty-first century with the capability of driving our daily lives; self-driving cars and disruptive products like virtual assistants Alexa and Siri are classic examples. Yet, AI that demonstrates superior intelligence and human-level consciousness is still evolving and is in the nascent stages. Peterson defines the four basic types of AI: Responsive AI, Limited Memory AI, Artificial General Intelligence, and Self-Aware AI.

i. *Responsive AI:* The base algorithm powering this first generation of AI is memory-free and responsive. When a certain input is received, the output is always the same. This type of AI-powered ML model is effective for easy classification and pattern recognition work. They can view huge amounts of data and produce reasonable results but cannot analyze situations where information requires incomplete or historical understanding (20).

ii. *Limited Memory AI:* The algorithms powering memory-limited machines work by replicating how the human brain works, and this variant is designed to mimic the connections of neurons in the human brain. This machine can address complex classification problems of deep learning and use historical data to make predictions. Tasks like autonomous driving and virtual assistants are perfect examples. Although the performance of Limited Memory AI is much superior to that of Responsive AI, it is widely classified as narrow intelligence. This is due to their innate tendency to lag human intelligence in several functional aspects. In machines powered by Limited Memory, AI needs a lot of training data to comprehend real-time issues in the form of apt examples. In this model, there is always a risk of abnormal or adverse specimens feeding the machines with false data leading to unwarranted outcomes (21).

iii. *Artificial General Intelligence:* It showcases generalized human cognitive abilities in software. When faced with unfamiliar tasks, AGI systems usually find relevant solutions. A typical AGI system is built to perform any task a human being can do. AGI is strong AI which contrasts with weak or narrow AI, usually applied to accomplish specific tasks. Overall, AGI is an intelligent system with cognitive computing capabilities, thanks to its ability to access and process big data at incredible speeds. The hallmarks of AGI are abstract thinking, background knowledge, common sense, cause and effect, and ability to transfer learning. Creativity, sensory perception, motor skills, and natural language comprehension are classic examples (22).

iv. *Self-Aware AI:* This is futuristic and highly disruptive in nature. It is an equivalent of AI becoming self-aware and achieving nirvana in the distant future. Right now, this type of AI exists only in theory holding lots of promise and possibilities. A typical self-aware AI is beyond human capabilities with independent intelligence, where people will probably need to negotiate terms with the artificial entity. The outcomes may vary in nature, and the future is wide open.

1.7 SIGNIFICANCE OF AI

AI is poised to play an indispensable role in the future as global data generation is tipped to touch 175 zettabytes (175 billion terabytes) by 2025, according to a 2018 report by research firm IDC. For businesses driven by informed decisions, access to big data is critical. Actionable information is the raw material for generating business intelligence with the power to optimize business activities. And this is where AI enables enterprises to take advantage of big data warehouses to drive core functionality. AI's ability to make meaningful predictions by negating human prejudice makes it a true game changer. Businesses deploying a high-quality cloud computing environment automatically enable AI applications. This affords organizations the computing capability to manage big data in a scalable and flexible architecture, empowering a wider range of users across the organization (23).

1.8 THE INFLUENCE OF AI ON THE ENTERPRISE DOMAIN

The value proposition of AI in the twenty-first century can be compared to the strategic value of electricity in the early twentieth century when electrification triggered the industrial revolution and enabled mass communications. As per Chris Bram, partner and director at Bain & Company, "Artificial intelligence is strategic because the scale, complexity, and dynamics of today's business are so great that people can no longer use it without artificial intelligence." The biggest influence AI will have on the business sector shortly will be in the form of automation and the ability of the workforce to complement the same. The increase in the AI workforce is expected to expand and surpass the profits generated by today's workplace automation tools. By analyzing huge amounts of data, AI not only automates the work environment but also creates the most efficient way to accomplish tasks in dynamic situations. AI has already optimized human activity in many cases. From helping physicians to diagnose accurately and enabling call center agents to render customer support more efficiently, AI has been delivering results across diverse domains in a sustained format. Regarding security, AI is used to respond to cyber-security threats and prioritize human attention automatically. Banks increasingly use AI to accelerate and support loan processing and regulatory compliance. On the flip side, AI can potentially eliminate a lot of human intervention and thereby the need for people (24). This is a big concern for operators.

1.9 THE ORGANIZATIONAL ADVANTAGES OF AI

The contemporary timeline is witnessing more and more enterprises leveraging AI to optimize their operations and output (Table 1.1). This is one of the most significant benefits of adopting AI in an enterprise environment. There are a host of other benefits offered by AI at both micro- and macro-levels to organizations pertaining to diverse domains and sizes (25).

- *Improve Customer Service:* AI's ability to accelerate and optimize customer service is one of the main potential benefits and was ranked second among AI objectives in a 2019 study by MIT Sloan Management and Boston Consulting.

TABLE 1.1
AI: The Timeline

Timeline	Artificial Intelligence
1940	The Enigma machine was decoded using AI during World War II
1950	Alan Turing releases a test to test machine intelligence
1955	John McArthy terms the phrase AI
1961	Unimate, the first industrial robot, was unveiled
1964	The first chatbot was invented by Joseph Wiezenbaum
1969	Shakey the Robot, the world's first general-purpose mobile robot, was unveiled
1995	Alice, the world's first chatbot, was introduced by Richard Wallace
1997	AI-powered gamer DeepBlue beats chess legend, Garry Kasparov
1998	The birth of Kismet, the first robot to be equipped with emotions
2002	Roomba, the world's first AI-powered vacuum cleaner, was introduced
2008	Voice recognition on I-phone and the birth of Siri
2011	IBM Watson, the question-answering machine, is unveiled
2014	Amazon introduces Alexa as a virtual assistant on all its devices
2016	Sophia the Robot becomes the first robot to receive citizenship (Saudi Arabia)
2020	A revolutionary tool for automated conversations, GTP-3 is introduced
2022	AI has become an indispensable part of workforces across domains worldwide
2023	AI adoption skyrocket by the rapid diffusion of models into all kinds of applications

- *Improve the Ability to Process Data in Real Time:* Businesses can automate and monitor processes almost instantly for better productivity. For example, manufacturing plants use image recognition software and ML models in their quality control processes to identify and address production problems.
- *Product Development:* AI can drastically shorten the new product development cycle, including the time between design, development, and commercial deployment, for a faster return on investment.
- *Optimal Standards:* Organizations using AI for tasks performed manually or by traditional thematic automation tools such as extraction, conversion, and reconciliation experience fewer errors by adhering to stringent compliance standards.
- *Advanced Talent Management:* Organizations deploying enterprise AI software can streamline the hiring process. Eliminating possible corporate biases, amplifying productivity by hiring the right fitment, etc. are some of the larger benefits. Advances in speech recognition and other NLP tools allow chatbots to provide personalized services to job seekers and employees.
- *Business Innovation and New Model Development:* E-commerce giants like Amazon, Airbnb, Uber, and others leverage AI to create new business models. Even mid- and small-size businesses are using AI to change their business models. The transition from traditional models to digital modes is happening fast due to AI (26).

1.10   RISKS OF AI

One of the biggest risks in using AI effectively in an organization is employee distrust. Many employees fear and distrust AI as they are unsure about the value of AI in the workplace. As per a recent Brookings Institution report on automation and AI, the impact of home and office automation via AI is such that almost 36 million jobs will soon be automated without the need for human intervention across various sectors in the near future. And by 2030, almost every profession will be affected by AI-based automation. On the brighter side, more than 500,000 jobs would soon be created for automation experts. By 2025, consultants predict that AI will create 2 million new jobs (27). Yet, the benefits of AI would not be realized without employee confidence and buy-in.

Elaborating about the same, Ammanat, MD of AI at Deloitte Consulting, told Tech Target, "I have seen cases where the algorithm has worked perfectly. But the workers are not trained and are not interested in using it." Consider an example of an AI method in a factory determining when a production machine should be shut down for service. "You can create a good AI solution—probably 99.998% accurate—and probably ask factory workers to stop. But if the end user does not trust the machine, which is not unusual, then AI fails," opined Ammanat. As AI models become more sophisticated, the ability to explain successfully how AI has reached its conclusion will become increasingly difficult for front-line employees who want to trust AI to make decisions. According to experts interviewed for a recent report on AI techniques, organizations should put the user first. Data processors should focus on providing relevant information to professionals, instead of going deeper into how the model works. For example, for ML models that predict hospitalization risk, physicians may need an explanation for the underlying medical reasons (28).

Factoring in all the critical components of AI, the broader risks associated with the domain can be classified as follows:

- *AI Error:* Although AI can eliminate human errors, problematic AI data errors can be caused by poor training data or algorithm errors. "Data errors can be dangerous due to the large number of transactions that AI systems usually process," says AI expert Brain of Bine & Company, a ship that processes millions of transactions a day.
- *Unethical and Unintentional Actions:* Companies must defend against unethical AI. For example, law firm executives are often aware of the discriminatory reports included in AI-based

risk prediction tools that some judges use to convict or acquit the accused. Corporations must be aware of the unintended consequences of using AI in their business decisions. Another example would be a grocery chain that uses AI to set competitive prices for other grocery stores. It is wise to charge more for food in poor areas with the low or absent competition. But it is also strategically important to remember about the grocery chain.

- *Violation of Critical Skills:* This is a trivial yet often neglected risk of AI deployment. After two AI-powered planes crashed with a Boeing 737 Max, some experts expressed concern that the pilots may have lost basic flight skills or at least the ability to use them. Since the aircraft relies on AI in its cockpit, this is an extreme case. But such occurrences can raise questions about the core competencies a business wants to retain among its employees.

1.11 CURRENT BUSINESS APPLICATION OF AI

A Google search for "using AI" yields millions of results, indicating many enterprises-related AI applications. Indeed, the AI-based industry is expanding from financial services—the primary recipient—to healthcare, education, marketing, and retail. All business sectors, including marketing, finance, and human resources, have embraced AI. It also includes several AI applications; natural language generation tools used in customer service, deep learning platforms used in autonomous driving, and oral recognition tools used by law enforcement agencies. Financial services are rapidly changing, thanks to how AI-integrated banks work and how they serve customers. Banking, financial services, and insurance (BFSI) players like Chase Bank, JP Morgan Chase, Bank of America, Wells Fargo, and other entities are leveraging the power of AI to improve their back-office operations, automate customer service, and create new opportunities. In the manufacturing sector, collaborative robots, also known as cobots, partner with humans in production lines and warehouses. The factory, a useful toolkit, deploys AI to forecast service demand. ML algorithms accurately determine purchasing behavior to predict product demand for production planning. The agriculture sector is also witnessing a paradigm shift in operations. The five trillion-dollar agribusiness uses AI to grow healthy crops, relieve pressure, control data, and perform other purposes. The document-driven legal industry uses AI to save time and improve customer service. Law firms study engineering to collect data and predict outcomes to describe data-related problems. They use computer vision to classify and separate documents and data. The education sector uses AI to automate tedious assessment processes and customize courses for specific needs. It is not an overstatement to say that ML can teach people and open the doors to personal learning. In the IT services sector, companies use NLP to process their service requests automatically. They apply ML to IT service management data to understand infrastructures and processes. From a broader perspective, IT companies use AI to optimize their IT services (29).

1.12 SPECIFIC BUSINESS APPLICATIONS OF AI ACROSS DIVERSE SECTORS

i. *E-commerce*

- *Personalized Shopping:* AI technology is leveraged to create recommendation engines through which marketers can engage better with customers. The recommendations are made after factoring in browsing history, preferences, and interests. It is a highly efficient customer relationship management (CRM) tool that helps foster customer loyalty toward the brand.
  - *AI-Powered Virtual Assistants:* Virtual shopping assistants like chatbots drastically improve the online shopping experience. NLP is deployed to mimic human conversation and to make it sound as personal as possible, and virtual assistants are built to have real-time engagement with customers, thus driving up sales and customer satisfaction levels.

- *Fraud Prevention:* Credit card frauds and fake online reviews are two of the most critical issues confronting e-commerce entities. Factoring in the usage patterns, AI helps tremendously reduce the possibility of credit card fraud. Since most customers depend on online reviews before making a buying decision, it is pertinent to eradicate fake reviews, and AI has the power to identify and weed out fake reviews.

ii. *Academics*

The education sector is undoubtedly mostly powered by humans, yet AI has slowly started making its presence felt in the academic sector as well. The slow yet steady transition of AI is driving up productivity among faculties. AI is helping the teaching staff concentrate more on students than on office and administration work.

- *Automated Administrative Tasks:* AI assists academicians with non-teaching tasks like automating customized messages to students, back-office tasks like filing, facilitating interactions with parents, feedback rollout, student enrolment, curriculum development, and admin tasks.
- *Smart Content Creation:* AI assists in creating a rich learning experience by generating and publishing audio–video summaries and comprehensive lesson plans. AI also facilitates digitizing content, including web/video lectures, conferences, and textbooks. Using AI, it is also possible to infuse different interfaces like animations and special effects for an interactive learning experience from a student perspective.
- *Voice Assistants:* Students can access extra learning material or get assistance through AI-powered Voice Assistants without the presence of a teacher in person. This also facilitates minimizing printing and stationary costs.
- *Personalized Learning:* Leveraging AI technologies, hyper-personalization tools can be deployed to monitor students' data thoroughly. It also helps gain a view of students' habits and performance. Generation of lesson plans, alerts, study guides, flash notes, revision notes, etc. is also possible (30).

iii. *Lifestyle*

- *Autonomous Vehicles:* The big-ticket automobile entities like Toyota, Audi, Volvo, and Tesla use ML to train computers to detect objects and avoid accidents. These computers can think and evolve like humans when driving in any environment.
- *Spam Filters:* The email integral to people's day-to-day lives is embedded with AI that filters out spam emails; these are redirected to spam or trash folders, allowing the user to see filtered content only. Market leader Gmail has reached a filtration capacity of approximately 99.9%, which is an achievement.
- *Facial Recognition:* Daily life devices like smartphones, laptops, and desktops use facial recognition techniques by deploying face filters; these filters detect and identify users' faces to grant secure access. Facial recognition is also a widely used AI application, even in high-security, sensitive areas across diverse industries.
- *Recommendation System:* Various platforms popular in daily life, like e-commerce, entertainment websites, social media, video-sharing platforms (like YouTube), etc. all use the recommendation system. It is deployed to capture user data and provide customized recommendations to amplify engagement. This AI application is deployed by almost all industries (31).

iv. *Navigation*

As per a study by the reputed MIT, GPS technology has the potential to provide users with accurate, timely, and specific information to improve safety drastically. The technology combines Convolutional Neural Network and Graph Neural Network; it simplifies navigation in traffic by automatically detecting the number of lanes and road types covered by obstructions on the roads. AI is deployed by Uber and many logistics firms to amplify operational efficiency, accurately analyze traffic, and optimize routes.

v. *Robotics*

Robotics is yet another area where AI applications are widely used. Robots powered by AI leverage real-time updates to instantly detect obstacles in path and preplan routes. It is used for various assignments, including carrying goods in hospitals, factories, and warehouses; cleaning offices and large equipment; and inventory management.

vi. *Human Resource Management*

Most organizations now use AI-based software to simplify the hiring process. It helps especially with blind hiring. HR managers can use ML software to analyze applications based on specific parameters. AI-drive systems scan aspirants' profiles to give recruiters an overview of the available talent pool.

vii. *Healthcare*

AI finds a wide array of diverse applications across the broader healthcare sector. The healthcare sector leverages AI applications to innovate state-of-the-art machines that detect ailments and identify malignant cells. AI can also help analyze chronic conditions using diagnostic and medical data of patients to deliver an early diagnosis. AI tends to synergize historical data and medical intelligence to discover new drugs, most of which can be lifesaving.

viii. *Agriculture*

AI is widely used in many topographies to identify defects and nutrient deficiencies in the soil. This is accomplished using a combination of computer vision, robotics, and ML applications. AI can also use the same combination to detect weeds in the crop. AI-powered robots are being deployed to harvest crops at a higher volume and a faster pace as compared to human labor. This shortens the process in the farm-to-shelf cycle.

ix. *Gaming*

AI applications have found prominence in the gaming sector as well. AI is leveraged to create smart, almost human-like NPCs to interact with the players. The technology is also widely deployed to predict human behavior; this is used to improve game design and testing. In 2014, the Alien: Isolation game initially announced that AI was utilized to stalk the player throughout the game. The game uses two AI-based systems: "Director AI" and "Alien AI"; the first one frequently knows the player's location, and the second is driven by sensors and behaviors and continuously hunts the player.

x. *Social Media*

- *Instagram:* On Instagram, AI factors in users' likes and the accounts followed to determine what posts should be shown on the user's explore tab.
- *Facebook:* In this scenario, AI is used along with a tool called DeepText. With this tool, Facebook can comprehend conversations better. This is used to translate posts from different languages automatically.
- *Twitter:* Twitter deploys AI for fraud detection, weeding out propaganda and hateful content. Twitter also leverages AI to recommend tweets that users might like based on the type of tweets they engage with.

xi. *Marketing*

AI-powered marketing can deliver precision-targeted and customized results aided by behavioral analysis and pattern recognition. AI also helps retarget potential consumers at the right time to deliver better results by reducing annoyance and building credibility. AI can be a useful tool in driving content marketing strategies in tune with a brand's style and tone of voice. It can also carry out routine tasks like performance assessments, campaign reports, and much more. Tools like chatbots, NLP, Natural Language Generation, and Natural Language Understanding efficiently analyze the user's preferred language and respond in a way humans do. AI gives users real-time personalizations based on behavioral patterns; the same data is used to formulate and optimize marketing campaigns with regional sensibilities.

xii. *Finance*

As per a recent report, 80% of banks worldwide acknowledge the benefits that can be accrued by deploying AI. Be it personal finance, corporate finance, or consumer finance, solutions formulated using AI can significantly enhance the quality of a wide range of financial services. For instance, customers looking for wealth management solutions can easily access relevant information through AI-driven SMS text messaging or online chat. AI can also identify changes in transaction patterns and other potential red flags; this can help detect fraud, which humans may miss. This can go a long way in saving enterprises and individuals from considerable losses. Besides fraud detection and task automation, AI can predict and calculate loan risks precisely.

1.13   EVOLUTION OF AI

The scope for using Enterprise AI is rapidly expanding. It has also evolved in the face of today's market uncertainty. While the enterprise sector is still battling the effects of the coronavirus epidemic, AI is helping the sector understand how to stay relevant and profitable. Quoting Arijit Sengupta, CEO of AI platform provider Aibel, "The most important uses are focused on scenario planning, proposal testing and hypothesis testing." It is important to seek feedback from end users at an indefinite time for any use. This is because end-user data knows what is still being discovered. The basic modeling needs to be more flexible and repetitive to get the most out of AI applications. Real-time empirical data is crucial for continuous analysis. For instance, forecast models may hire salespeople who will also become key players in feeding the model. Increasing investment in AI and ML in real estate and supply chain management, such as heavy machinery, is a classic case (32).

1.14   IMPLEMENTATION OF AI IN ORGANIZATIONS

Recent research on adopting AI in companies shows that the number of AI applications is increasing. In the recent past, Gartner had rightly predicted that by 2022 the average number of AI projects per company would increase from 35, which was 10 times higher than the then average. IBM's "Global AI Adoption Index 2021," conducted by Morning Consult on behalf of IBM, shows that while most AI adoption remained unchanged between 2020 and 2021, a "significant investment plan" study found that about three-quarters of companies are already deploying AI or exploring the possibility of the same. Most respondents also stated that their companies are accelerating AI adoption because of COVID-19 (33). Other significant findings include the following.

- 91% of AI users said that strong and intelligent AI is important for businesses as it is important for management to explain how AI makes decisions.
- About 90% of IT professionals said accessing information from anywhere is the key to extending AI adoption, as AI projects can be launched wherever.
- NLP is at the forefront of recent adoption; today, about half of businesses use NLP-based applications, and 25% said they plan to start using NLP technology within the next 12 months.

1.15   STEPS IN IMPLEMENTING AI IN ORGANIZATIONS

AI has many forms: ML, in-depth learning, predictive analysis, NLP, computer vision, and automation. Companies need to address issues related to people, processes, and technologies to benefit from the multitude of AI technologies (34). The ten most significant steps in implementing AI-based technologies in enterprises are as follows:

i. Ensure fluency of information.
ii. Identify key business incentives for AI.

iii. Determine the scope of opportunity.
iv. Evaluate your internal abilities.
v. Identify suitable candidates.
vi. Design AI pilot project.
vii. Build a basic understanding.
viii. Schedule periodic scale.
ix. Identify general AI power to bring maturity.
x. Continuously improve AI model and process.

However, like every new technology, codes of conduct are still being written. AI industry leaders insist that experimental ideas will yield better results.

1.16　NEW ACHIEVEMENTS IN THE FIELD OF AI

- *Games:* Machines need to consider many possible positions, such as chess, river crossings, N-Queen problems, etc.
- *Natural Language Processing:* Interact with a computer that understands.
- *Expert Systems:* Provides explanations to machine or software users.
- *Computer Vision System:* The system understands, interprets, and describes the visual input to the computer input on the computer.
- *Intelligent Robot:* The robot can follow the instructions given by humans.

1.17　MAJOR GOALS OF ENTERPRISE-GRADE AI

- Computer scan
- Knowledge-based feedback
- Machine learning
- Planning
- Processing of natural language

1.18　IBM WATSON ROBOTICS

Watson is an IBM supercomputer that combines AI and complex programming to act as a fully responsive tool. It deserves special mention as part of this project purely due to its status as the pioneer in this domain. The supercomputer is named in honor of IBM founder Thomas SJ Watson. IBM Watson was at the beginning of a new computing era when it was created (35). Recently, Watson's limitations have been increased, and the way Watson works has moved to a new, distorted delivery model (IBM Cloud Watson) with ML capabilities and updated hardware. It is no longer a question that responds to organized search engines outside of participating in questions and answers; now it is "see," "hear," "read," "speak," "try," "translate," "learn," "I know," and "I am certified."

1.19　AI TRENDS

A host of worldwide vendors, governments, and research institutes are working to create AI. Simultaneously, start-ups in hardware are developing new approaches in the organization of memory, computing, and networking, which could change the functionality of leading organizations and the method of deploying AI algorithms. At least one vendor has begun testing a single iPad-sized chip that can transfer data faster than existing AI chips. Conventional wisdom (such as the superiority of GPUs over processors in handling AI workloads) is also being questioned by AI researchers. It is being proven that processors are common hardware: They are everywhere and cheaper than GPUs, capable of being game changers.

Educators and industry scientists in software are pushing the boundaries of today's AI applications. Intense efforts are made to create a sensitive machine that competes with human intelligence. Symbolic AI Proponents is one such method based on problem presentation, logic, and high-level symbolic search. The goal of joining forces with proponents of data-intensive neural networks is to develop an AI that better captures the structure and causes of the face of symbolic AI. It can recognize images created by deep neural networks and process natural language. This neural symbolic mechanism will enable machines to reason with what they see. It represents a milestone in the development of AI.

Modeling neural symbols is now one of the most exciting areas of AI, as per Brendan Lake, an assistant professor of psychology and data science at New York University. AI developments that seemed daunting a few years ago are becoming institutional sellers who combine these advances with commercial products. These efforts are being pioneered by tech giants like Google, FB, Tesla, etc. Other significant trends include the following:

- Automated ML that improves data identification and automated optimization of neural network architectures.
- Multi-learning AI that supports formats such as text, vision, speech, and IoT sensor data in a more integrated machine format.
- Mini ML is a model of AI and ML that runs on limited hardware devices, such as microcontrollers that control machines, refrigerators, and service meters.
- Concept-driven AI design AIK is trained to take on special roles in fashion, architecture, design, and other creative skills.
- New AI models, such as DALL·E, can generate a design concept for new things.
- Quantum ML is powered by quantum computing resources and simulators created by tech entities like Microsoft, Amazon, and IBM Cloud with pathbreaking abilities.

1.20   AI: THE FUTURE IS HERE

People have long feared the rise of cyborgs—their creations evolving into smarter and more intelligent versions of humans. Yet, AI and ML are being embraced by an increasing number of people and enterprises in daily lives and business functionality. This trend is rapidly changing our world, and we are amid the fourth industrial revolution officially. Whether or not our future generations are being outsmarted by cyborgs lording over the planet is a million-dollar question; AI is surely enriching our lives and businesses. It is a reality that is fast taking shape.

- *Creating New Career Avenues:* AI will alter the workforce in times to come. The scary view of AI as a job killer is just the flip side. While 75 million jobs are tipped to disappear, around 1.3 million more engaging, less repetitive new ones are forecasted to be created. AI is evolving as an opportunity for workers to focus on specific parts of their jobs that may drive high levels of job satisfaction.
- *Closing Linguistic Gaps:* Be it teaching new languages in a customized way or translating speech and text in real time, AI-based language tools (Duolingo, Skype) are comprehensively closing linguistic and sociocultural divides across workplaces, classrooms, and drawing rooms. Though digital translation services are still in the nascent stages of evolution, it can be safely stated that they provide a means of understanding that might be otherwise impossible.
- *Transforming Governance:* Imagine a world with lesser paperwork, quicker turnaround time, and highly responsive governance. Well, AI has the power to achieve all these and more, including smarter public administration and transparent governance. However, this comes with both pros and cons that need thorough comprehension and evaluation. In a

hypothetical scenario, tools like gamification and role-playing could be leveraged by public servants to assess complex cases and formulate better solutions. This would pave the way for fully and truly absorbing the future role of autonomous systems in public governance.

- *Healthcare Delivery:* AI has the power to make healthcare more accessible and more affordable. Interestingly, an AI-powered chatbot offering symptom checking and fast access to physicians (if needed) can provide advice to more people in remote locations. This way, patients can access more accurate, affordable, safe, and convenient healthcare solutions within very short periods, and AI helps healthcare providers save on overhead costs by deploying efficient chatbots.
- *Art Creation:* Computational creativity is radically altering the nature of art. Software is now more than a tool; it is evolving into a creative collaborator, seamlessly synergizing the capabilities of a computer scientist and an artist. It is more like an exhibition space metamorphosing into a lab where art is an expression of science, with the artist as the researcher.

1.21 INTERESTING AI FACTS AND FIGURES

- As per Statista, income generated from the AI software market globally is predicted to reach 126 billion dollars by the year 2025.
- As per a recent study by Gartner, 37% of organizations have already implemented AI in some form. The percentage of organizations deploying AI grew 270% over the past four years.
- As per Servion Global Solutions, by 2025, 95% of customer interactions will be helmed by AI.
- A recent 2020 study by Statista shows that the worldwide AI software market is tipped to grow approximately 54% YoY; it is predicted to reach a forecast size of US$22.6 billion.

1.22 CONCLUSION

As human beings, we have always been fascinated by technology along with fictional changes, and we are now experiencing some of the greatest advances in our history. AI has become the next big challenge in technology. Businesses and corporations around the world are pushing for innovation in ML. AI influences not only the future of every industry and every person but also serves as a major driver for emerging technologies such as big data, robots, and the IoTs. The enterprise sector will continue to be a technology innovator in the future. As a result, well-trained and certified professionals have many opportunities for successful careers. As these technologies continue to evolve, there will be a greater influence on the social environment and quality of life. With advances in AI like face recognition, healthcare, chatbots, and more, the future belongs to creative intelligence. Virtual assistants have already entered everyday life, saving time and effort. Collective with this is just the beginning, with a lot to come. One of the features that sets us apart from millions of years of history on earth is our reliance on tools and our commitment to improving the tools we invent. Once we understand how AI works, AI tools will inevitably become more intelligent. It has to do with the future of everything we do. Indeed, the strangeness of AI in the workplace is nothing more than a hammer or a plow. However, the quality of AI tools differs from all other tools in the past. We can talk to them, and they will respond. They quickly enter our space, answer our questions, solve our problems, and make our life easier. The big difference between human intelligence and AI is disappearing fast. This uncertainty between human intelligence and AI comes from other technological trends, which inadvertently shuffles AI. Finally, smart growth can be two-way in the future, making our machines and us smarter. AI is all about creating a win-win situation (36)!

1.23   MEMORABLE QUOTES ON AI BY EMINENT PERSONALITIES (37)

- *Elon Musk:* "AI is neither good nor evil. It's a tool. It's a technology for us to use."
- *Oren Etzioni:* "I think there should be some regulations on AI."
- *Andrew Ng:* "We can build a much brighter future where humans are relieved of menial work using AI capabilities."
- *Fei-Fei Li:* "AI is everywhere. It's not a that big, scary thing in the future. AI is here with us."
- *Sam Altman:* "AI will probably most likely lead to the end of the world, but in the meantime, there'll be great companies."
- *Lisa Joy:* "And nowadays, the idea of AI is not really science fiction anymore—it's just science fact."
- *Vernor Vinge:* "When people speak of creating superhumanly intelligent beings, they usually imagine an AI project."
- *Rana el Kaliouby:* "The real problem is not the existential threat of AI. Instead, it is in the development of ethical AI systems."
- *Vivienne Ming:* "AI might be a powerful technology, but things won't get better simply by adding AI."
- *Reid Hoffman:* "If you could train an AI to be a Buddhist, it would probably be pretty good."
- *Marc Benioff:* "I think a lot of people don't understand how deep AI already is in so many things."
- *Claude Shannon:* "I visualize a time when we will be to robots what dogs are to humans, and I'm rooting for the machines."
- *Gray Scott:* "The real question is, when will we draft an artificial intelligence bill of rights? What will that consist of? And who will get to decide that?"

REFERENCES

1. Feijóo C, Kwon Y, Bauer JM, Bohlin E, Howell B, Jain R, et al. Harnessing artificial intelligence (AI) to increase wellbeing for all: The case for a new technology diplomacy. Telecommunications Policy. 2020;44(6):101988.
2. Nouri J, Zhang L, Mannila L, Norén E. Development of computational thinking, digital competence and 21st century skills when learning programming in K-9. Education Inquiry. 2020;11(1):1–17.
3. Chatterjee R. Fundamental concepts of artificial intelligence and its applications. Journal of Mathematical Problems, Equations and Statistics. 2020;01:13–24.
4. Oliveira E. Beneficial AI? Fight for it! International Journal of Computers and Their Applications. 2017;24:169.
5. Russell S. A binary approach. Ethics of Artificial Intelligence: Oxford University Press; 2020. p. 327.
6. Domingos P. The Master Algorithm: How the Quest for the Ultimate Learning Machine Will Remake Our World: Basic Books, Inc.; 2018.
7. Brooks R. The Seven Deadly Sins of AI Predictions. MIT Technology Review; 2017. Available from: https://www.technologyreview.com/2017/10/06/241837/the-seven-deadly-sins-of-ai-predictions/.
8. Oliveira E. Beneficial AI: The next battlefield. Journal of Innovation Management. 2018;5:6.
9. Yang X, Wang Y, Byrne R, Schneider G, Yang S. Concepts of artificial intelligence for computer-assisted drug discovery. Chemical Reviews. 2019;119(18):10520–94.
10. Pei Q, Luo Y, Chen Y, Li J, Xie D, Ye T. Artificial intelligence in clinical applications for lung cancer: Diagnosis, treatment and prognosis. Clinical Chemistry and Laboratory Medicine. 2022;60(12):1974–83.
11. Davenport T, Kalakota R. The potential for artificial intelligence in healthcare. Future Healthcare Journal. 2019;6(2):94–8.
12. Younes OS. A secure DHCP protocol to mitigate LAN attacks. Journal of Computer and Communications. 2016;4:39–50.
13. Lavin A, Gilligan-Lee CM, Visnjic A, Ganju S, Newman D, Ganguly S, et al. Technology readiness levels for machine learning systems. Nature Communications. 2022;13(1):6039.
14. Coursera. 3 Types of Machine Learning You Should Know: Coursera; 2022. Available from: https://coursera.org/share/e93159a6dd2ed098badb10bcd0973b93.

15. Taeb M, Chi H. Comparison of deepfake detection techniques through deep learning. Journal of Cybersecurity and Privacy [Internet]. 2022;2(1):89–106.

16. Batko K, Ślęzak A. The use of big data analytics in healthcare. Journal of Big Data. 2022;9(1):3.

17. Burns E. What is artificial intelligence (AI)? TechTarget; 2023. Available from: https://www.tech-target.com/searchenterpriseai/definition/AI-Artificial-Intelligence#:~:text=AI%20programming%20focuses%20on%20three,the%20data%20into%20actionable%20information.

18. Wang D, Prabhat S, Sambasivan N, editors. Whose AI dream? In search of the aspiration in data annotation. Proceedings of the 2022 CHI Conference on Human Factors in Computing Systems; 2022.

19. Arkoudas K, Bringsjord S. Philosophical foundations. The Cambridge Handbook of Artificial Intelligence: Cambridge University Press; 2014:34–63.

20. Khamparia A, Gupta D, Khanna A, Balas VE. Biomedical Data Analysis and Processing Using Explainable (XAI) and Responsive Artificial Intelligence (RAI): Springer; 2022.

21. Ashrafuzzaman M, Das S, Chakhchoukh Y, Shiva S, Sheldon FT. Detecting stealthy false data injection attacks in the smart grid using ensemble-based machine learning. Computers & Security. 2020;97:101994.

22. Goertzel B. Artificial general intelligence: Concept, state of the art, and future prospects. Journal of Artificial General Intelligence. 2014;5(1):1.

23. Stylos N, Zwiegelaar J, Buhalis D. Big data empowered agility for dynamic, volatile, and time-sensitive service industries: The case of tourism sector. International Journal of Contemporary Hospitality Management. 2021;33(3):1015–36.

24. Brougham D, Haar J. Smart technology, artificial intelligence, robotics, and algorithms (STARA): Employees' perceptions of our future workplace. Journal of Management & Organization. 2018;24(2):239–57.

25. Weiss G. Multiagent Systems: A Modern Approach to Distributed Artificial Intelligence: MIT press; 1999.

26. Wu B, Widanage WD, Yang S, Liu X. Battery digital twins: Perspectives on the fusion of models, data and artificial intelligence for smart battery management systems. Energy and AI. 2020;1:100016.

27. Rajnai Z, Kocsis I, editors. Labor market risks of industry 4.0, digitization, robots and AI. 2017 IEEE 15th International Symposium on Intelligent Systems and Informatics (SISY); 2017: IEEE.

28. Beaulieu-Jones BK, Yuan W, Brat GA, Beam AL, Weber G, Ruffin M, et al. Machine learning for patient risk stratification: Standing on, or looking over, the shoulders of clinicians? NPJ Digital Medicine. 2021;4(1):62.

29. Sorescu A, Schreier M. Innovation in the digital economy: A broader view of its scope, antecedents, and consequences. Journal of the Academy of Marketing Science. 2021;49(4):627–31.

30. Benson A, Odera F. Selection and use of media in teaching Kiswahili language in secondary schools in Kenya. International Journal of Information and Communication Technology Research. 2013;3(1):12–18.

31. Tyagi AK, Fernandez TF, Mishra S, Kumari S, editors. Intelligent automation systems at the core of industry 4.0. Intelligent Systems Design and Applications: 20th International Conference on Intelligent Systems Design and Applications (ISDA 2020), December 12–15, 2020; 2021: Springer.

32. Abioye SO, Oyedele LO, Akanbi L, Ajayi A, Delgado JMD, Bilal M, et al. Artificial intelligence in the construction industry: A review of present status, opportunities and future challenges. Journal of Building Engineering. 2021;44:103299.

33. Henke N, Puri A, Saleh T. Accelerating Analytics to Navigate COVID-19 and the Next Normal: Mckinsey; 2020.

34. Jarrahi MH. Artificial intelligence and the future of work: Human–AI symbiosis in organizational decision making. Business Horizons. 2018;61(4):577–86.

35. High R. The era of cognitive systems: An inside look at IBM Watson and how it works. IBM Corporation, Redbooks. 2012;1:16.

36. Weibin P, Liuqing F, Xiaojing L. Digital governance for smart city and future community building: From concept to application. Smart Cities for Sustainable Development: Springer; 2022. pp. 41–67.

37. Bushe GR, Kassam AF. When is appreciative inquiry transformational? A meta-case analysis. The Journal of Applied Behavioral Science. 2005;41(2):161–81.

2 An Overview of Artificial Intelligence-driven Pharmaceutical Functionality

Hanan Fahad Alharbi
Princess Nourah bint Abdul Rahman University, Riyadh, Saudi Arabia

Mullaicharam Bhupathyraaj
College of Pharmacy, National University of
Science and Technology, Muscat, Oman

Kiruba Mohandoss
Sri Ramachandra Institute of Higher Education and
Research, Chennai, Tamil Nadu, India

Leena Chacko
Meso Scale Diagnostics LLC, Rockville, Maryland, USA

K. Reeta Vijaya Rani
Surya School of Pharmacy, Vikravandi, Tamil Nadu, India

2.1 AN INNOVATIVE INTRODUCTION TO ARTIFICIAL INTELLIGENCE (AI)

As per Oxford Dictionary (2021), the *formal explanation (definition) of AI* is as follows: "The theory and development of computer systems which can perform tasks that normally require human intelligence, such as visual perception, speech recognition, decision-making, and translation between languages." AI has a long history with several iterations in specific areas of interest and general research direction. Big data, affordable cutting-edge infrastructure, and futuristic technologies combine to create a unique synergy facilitating the deployment of AI in wide areas of application across the various sectors (1). The advent of AI to drive functional optimization across diverse domains has been rapid and is being embraced as a new age operational framework. Similarly, the implementation of AI across vital operations in pharmaceutical and healthcare is fast gaining traction driving a vast range of functions, including diagnostics, drug discovery, drug development, process design, and drug manufacturing (2).

2.2 THE CORE OBJECTIVE

AI is a subfield of computer science that has managed to evolve into a problem-solving science with widespread applications across diverse domains. The main objective of AI is to identify useful information-processing problems and provide an intangible account of how to solve them efficiently. This is a systematic method which relates to a hypothesis in mathematics (3). AI can be broadly described as a field that deals with the design and implementation of algorithms for data analysis, learning, and interpretation. AI covers a wide range of statistical and machine learning

DOI: 10.1201/9781003343981-2

(ML), pattern recognition, clustering, as well as similarity-based methods (4). AI is a fast-emerging technology that has applications in a wide range of areas of life and industry. In recent years, the pharmaceutical industry has discovered novel and innovative ways to use this powerful technology to help solve some of the industry's most pressing problems. In the pharmaceutical industry, AI refers to the use of automated algorithms to perform tasks that have traditionally relied on human intelligence (5). In the last five years, the use of AI in the pharmaceutical and biotech industries has transformed the innovative methods by which the scientists develop new drugs to combat disease and more. Over the last few decades, there has been rapid activity in the development of innovative systems for the targeted delivery of therapeutics with maximum efficiency and minimal side effects (6). Controlled drug delivery and tackling the challenges associated with traditional drug delivery systems (such as systemic toxicity, narrow therapeutic index, and dose adjustment in long-term therapy) have been the subject of intensive research. Furthermore, when it comes to controlled drug delivery, the use of micro-fabrication technology holds immense potential and promise (7).

By utilizing technological advancements in AI, the pharmaceutical industry can accelerate its development. The most recent technological trends pertaining to pharmaceutical sector point to increased deployment of computer systems capable of performing tasks which otherwise require human intelligence (visual processing, speech recognition, decision-making, and language translation) (8). According to the International Business Machines Corporation (IBM), the entire healthcare domain has at its disposal immense amounts of big data. With the availability of so much data in this domain, AI can be of great assistance in scrutinizing it and presenting results that will aid in precise decision-making, saving human effort, time, and money, thus improving healthcare and helping save lives (9).

2.3 AI FOR PHARMACEUTICAL WORLD

Historical data pertaining to the pharmaceutical field proves the efficacy of AI in effectively reducing the time and cost of drug discovery, development, and distribution. Before the advent of AI, the whole process of drug discovery was a costly and time-consuming affair, often spanning over many years (10). The highly complex nature of intricate biological systems further added layers to the drug discovery and developmental process. Drug discovery process involves initially identifying a molecule to modulate a particular step in a specific biochemical trial (11). Detailed background study and the effect of the particular molecule on the other components of the biological system are vital in developing any new drug using a specific molecule. Preclinical or lab studies may find many compounds to perform exceptionally well in a controlled environment. But it may not obtain buy-in from humans due to their antagonistic reactions (adverse drug reactions and allergic manifestation) (12). The sheer number of changes in the biomolecular structures and the potential capability of the number of molecules to transform into therapeutic drugs are amazing and awe inspiring (13). Recurrent tests with these molecules are time- and cost-consuming affairs warranting high level of skills and expertise. The process may take time to tune with the behavior of the molecular structure. ML and natural language processing (NLP) opens a wide number of possibilities leading to rapid solutions replete with trained data as per Figures 2.1 and 2.2. The inevitability of multilayer perceptron (MLP) along with NLP in pharmaceutical research and development (R&D) could accelerate critical processes seamlessly (14). The proven efficacy of AI in dealing with big data can be a harbinger of change in the pharmaceutical field. An efficient AI system is built to connect with the input replete with relevant data points as provided and from the group/set of inputs; it could point to precise candidate molecules based upon the set criteria. Developing a protein from a group of small molecules based upon a specific modulate function remains the overarching goal of drug delivery (15). The process may often be challenging without an appropriate hint or lead. Knowledge pertaining to medicinal properties of naturally occurring compounds gained or transferred over hundreds or thousands of years is found to be effective while pairing it with AI. The medicinal properties of various substances are mostly found in nature and are currently being used in the development of modern drugs (16). On the flip side, "De Novo" (newly developed drugs) have come into the world

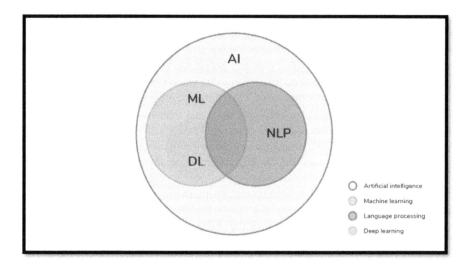

FIGURE 2.1 Machine learning (ML) and natural language processing (NLP) are subsets of AI that are often used together to achieve specific results. (Available at **https://athenatech.te**ch/f/ai-machine-learning-ml-and-natural-language-processing-nlp.)

because of the natural unavailability of certain molecules. The constant need for discovering new molecules which lead to the betterment of humankind is yet another factor accelerating the "De Novo" segment's growth (17). However, the whole exercise is a cumbersome process, wherein modern pharmaceutical firms leverage newer strategic and tactical methods of assessing a volume of molecules of the desired target (18). Despite the cutting-edge system for the evaluation and selection of the specifically identified targets, many times the outcomes are below par. This is mostly due to the unpredicted complex nature of the biological systems along with the adverse drug reactions, hypersensitivity reaction, or interactions of the molecules; very often they show positive interactions in the earlier stages but fail to provide a positive reaction in later stages in various regions of the biological system (19). Such issues may not always be there until drug development has undergone a huge volume of tests which consume high time and costs without yielding a feasible result (20). Various methods used in pharmaceutical sector are as follows:

- Advanced techniques like molecular analysis ("Machine—Vision") in AI use images in development of a system capable of predicting the specific molecules that might be effective and suitable for biological targets; this could also accelerate the drug discovery process (21).

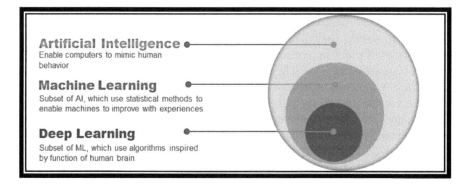

FIGURE 2.2 AI vs machine learning vs deep learning: understanding the differences. (Available at https://artificialintelligencefutureofworld.Blogspot.com/2021/01/what-is-ai-ml-and-dl-what-you-will-what.Html.)

- AI simulations with respect to chemical interactions can also be performed to evaluate the efficacy of a drug in testing a specific treatment. The current scenario demands the feasibility of choosing AI as a rapid medium of computational intelligence in the identification of appropriate vaccines for viral contagions (22).
- AI-based systems can identify and classify deadly pandemics like Ebola or Zika and rapidly mutating HIV viruses (23).
- Researchers in pharmaceutical companies have been using AI to understand the deep chemistry of drug interactions. Future uses of AI include study of biological systems' effect of a drug on a patient's tissues (e.g., how a chemical and a specific protein may interact). AI uses pharmaceutical inputs along with "Machine-Vision"-image analytics, to effectively produce a drug within a specific time and cost line by identifying the best-suited molecule that could positively act on humans (24).
- AI replete with ML and NLP enables various stages of drug development and could be a game changer in the field of pharma (25).

2.4   APPLICATION OF AI IN PHARMACEUTICAL RESEARCH IN DRUG FORMULATION

- *Controlled-release Tablets:* Initial research work using neural networks for the modeling of pharmaceutical formulations was held at the University of Cincinnati (OH, USA) by Hussain and coworkers (26). Via numerous studies, these researchers were able to model an *in vitro* release feature of a wide range of drugs that dispersed in matrices prepared from innumerable hydrophilic polymers (27). It has been analyzed that for all these cases, single-layered neural networks were able to provide acceptable performance in the predictive analysis of drug release (28). General analysis within the limited data proved that the neural networks were able to offer comparable predictions through statistical intervention. However, the scope of data changes will have an impact on the predictions and will not be stable unless it has been properly trained for the neural networks (29). The solutions to some of the formulations in drug discovery depend on optimized formulations. As the role of optimized algorithms for drug formulations using some of the best techniques like genetic algorithm, particle swarm optimizations, etc. was lacking and unknown (30), the scientists consider this powerful computer-aided formulation design using neural networks for future research. Researchers from KRKA d.d. (Smerjeska, Slovenia) and the University of Ljubljana (Slovenia) made some recent studies in the formulation of diclofenac sodium from a matrix tablet prepared via acetyl alcohol using neural networks in the prediction of the rate of drug release (31). They also made attempts to carry out optimization in the formulation using 2D and 3D response surface analysis. The outcomes showcase nonlinear relationships between the number of components used in the drug formulation and the drug release rate, signifying the flexibility of multiple formulation productions with similar drug release profiles (32).
- *Immediate-release Tablets:* Research on immediate-release tablets emerged at the beginning of 2018 with two studies, at the University of Marmara (Turkey) and the University of Cincinnati by Turkoglu and colleagues (33). Knowing the strengths of neural networks, they applied them to model tablet formulations of hydrochlorothiazide along with the support of a statistical approach. Three-dimensional plots for massing time, compression pressure, and crushing strength or drug release were generated using neural networks. Maximizing tablet strength or selecting the best lubricant to react with depends upon the massing time and compression pressure in drug formulation (34). Many drifts have been observed, but optimal formulations were not yet finalized. Some of the trends observed seem to show results that are close to those observed using statistical methodologies (35).

The networks created were used to generate 3D plots of massing time, compression pressure, and crushing strength (or drug release), to maximize tablet strength or choose the appropriate lubricant (36).

• *Product Development:* There's a lot of scope in pharmaceutical product development process, warranting multivariate optimization problems. The process needs optimization for the variables in the formulation of drugs (37). Artificial neural networks (ANNs) have the ability to analyze and generalize the data provided to them. Though the process in its initial stage demands a set of training actions, these characteristics enable them to be one of the most efficient in problem-solving systems (38). These features are encompassed to drive drug formulation in the development stage. ANNs showcased their ability to fit into the model and are enriched with the prediction capabilities in drug development in the form of solid dosage with the impact of multiple factors such as formulation factors, compression parameters, etc. on tablet properties, like dissolution (39). ANN's capability to provide a tool useful for the development of a microemulsion-based drug delivery system could provide a minimal experimental effort (40). The current research highlights the ability of ANNs to have a prediction-based effect in multidimensional research areas. The impact of ANN's predictive nature was used in predicting the phase behavior of quaternary microemulsion-forming systems, which comprise water, oil, and other two surfactants. Simulation of aerosol behavior (by the view of engaging this methodology for design and evaluation of pulmonary drug delivery system) found its path through ANNs (41). However, limitations are there with ANNs when the decision-making becomes too complex. Fuzzy logic, as the term states "fuzzy," could provide a solution that could define a problem solution in between the space of "yes" or "no" (42). A combination of fuzzy logic with decision-making and controlling along with neural networks makes it a powerful tool to solve unsolved problems (43). The combination will be powerful and offers more flexibility when fuzzy could take the data and decide on the conditions "if... so... then."

2.5 APPLICATIONS OF AI IN THE PHARMA AND HEALTHCARE INDUSTRY

• *The Road Ahead:* Success stories of AI in medical decision-making for the diagnosis of disease and classification are now on news every day. AI is a broader area that could strengthen the research in the direction of design with applications of various ANN-based algorithms for analysis, study, and interpretation of big data. AI is integrated with multidimensional branches with natural learning, ML (supervised, unsupervised), deep learning, and many more with the combination of soft computing or evolutionary computing or fuzzy logics/fuzzy controller modeling to be collectively called "Computational Intelligence." AI in pharmaceutics has a lot to accomplish with targets set for the drug developmental and drug delivery phases. AI has the potential to cut short the time-consuming testing process, eliminate unwanted cost implications in drug development, and provide an optimal drug formulation process (44).

• *Current and Future Role of AI in Shaping Pharmaceutical-Technology:* While AI grips multidimensional applications, its real value lies in the momentum gained through its best combination. The use of AI is expected to support the medical and pharmaceutical sectors with high ethics in its operational nature. As AI grows and expands with its branches reaching out to various programming sessions like NLP and MLP in the health sector, there are more angles waiting to be discovered:

  • In patient care and management—application of AI in tracking the health checkups of a patient along with a telemedicine focus (45).
  • In supporting the neuroimaging technicians to identify and classify diseases (46).
  • To accelerate the design of optimum drug formulation.
  • To identify critical cases from a hospital database and provide proper directions for healthcare delivery.

AI no doubt is already essaying an indispensable role in the functionality of the broader pharmaceutical spectrum. The coming years will witness amplified visibility of AI in almost every critical aspect of the pharma domain (47). Increasing dependence of the sector on AI for optimized outcomes is going to the future trend. In fact, AI is all set to drastically reduce human intervention in multiple aspects of the pharma domain's operational dynamics. Compared to the past, the use of AI in pharmaceutical technology has grown manifold over the years due to its ability to save time and resources (48). AI is gaining traction as a tool that provides insights into the diverse dynamics involved in both formulations and process parameters. Thanks to its versatility, AI has already evolved into a problem-solving science with numerous applications across a wide range of verticals within the broader pharma ambit; and the future promises more (49). The focus is on the use of AI in various sectors of the pharmaceutical industry, including but not limited to drug discovery and development, drug repurposing, improving pharmaceutical productivity, and clinical trials and continuum. The baseline is about drastically reducing the workload while meeting targets in the shortest possible lead time and this is simply because the future belongs to AI. However, the immediate future is all about the indispensable role of AI in mission critical processes like drug discovery, development, and delivery.

2.6 APPLICATION OF AI IN DRUG DELIVERY SYSTEMS

Creating a commercial product, whether it is a relatively simple formulation (capsule, tablet, oral liquid) or a controlled-release formulation (an implant), is always a time-consuming and complex issue. Overall, an initial formulation of one or more drugs mixed with various ingredients (excipients) is prepared; and as development progresses, the selection of these (replete with their levels), sometimes including the manufacturing process, is started as per configurations which are a result of significant, time-consuming experimental work. These iterations generate a large amount of data, which is difficult to process and understand. In reality, the formulator must work in a multi-dimensional design space that is nearly impossible to conceptualize. Statistics have been used as one approach to the situation till date (50). This approach has the advantage of producing models that are clearly expressed and have associated confidence factors. However, statistical approaches quickly become unwieldy for more than three or four inputs; so, the formulator is compelled to oversimplify the problem (for example, limiting a study to three input variables) to characterize it.

As shown in Figure 2.3, the microchip device has been developed to deliver a controlled pulsatile release of the polypeptide leuprolide from discrete, individually addressable reservoirs implanted

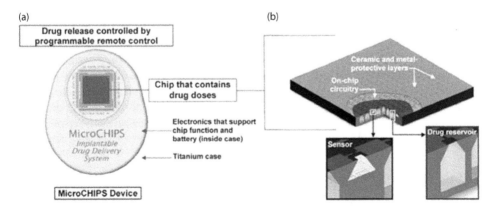

FIGURE 2.3 Microchip-based device for electronic format-controlled drug release. (a) The microchip is housed in a biocompatible case that also houses the electronic components, transmitter, and energy source for wireless technology. (b) Reservoir array for sensing applications or drug delivery. Each reservoir on the microchip contains 25 g leuprolide. The drug can be unveiled from the implantable and configurable device at monthly or weekly durations (51). (From ref. (52).)

in a canine model for nearly six months. This provides constant total drug concentration in the systemic circulation.

2.7 THE ADVENT OF NEURAL NETWORKS

Neural networks are mathematical constructs that can "learn" relationships within data without the user's prior knowledge. They provide superior solutions when compared to the statistics-based approach (53). The neural network makes no presumptions about the conceptual model of the relationships; instead, it generates and evaluates a variety of models to find the one that best fits the experimental data provided to it (54). As a result, ANNs are increasingly being used to model complex behavior in problems such as pharmaceutical formulation and processing. Neural networks, like humans, learn directly from input data. There are two types of learning algorithms. Unsupervised learning, in which the network is given input data and learns to recognize patterns in the data; this helps organize large amounts of data into fewer clusters. In supervised learning, the network is presented with a series of matching inputs and outputs, which are analogous to "teaching" the network; as a result, it understands the relationships linking the inputs to the outputs. Supervised learning has proven particularly useful in formulation, where the goal is to establish cause-and-effect relationships between inputs (ingredients and processing conditions) and outputs (measured properties). The networks created were used to generate three-dimensional plots of massing time, compression pressure, and crushing strength (or drug release) to maximize tablet strength or choose the appropriate lubricant (55).

2.8 AI IN DRUG FORMULATION

During this period, though trends were observed, no optimal formulations were provided. The trends resembled those produced by statistical procedures. Genetic algorithms were used to generate comparable neural network models, which were then optimized. The optimum formulation was discovered to be dependent on the constraints applied to the ingredient levels; these were used in formulation as well as in defining the relative importance placed on the output parameters. Disintegration time was sacrificed to achieve high tablet strength and low friability. Lactose was the recommended diluent in all cases and a fluidized bed was the preferred granulating technology (56).

2.9 AI IN PRODUCT DEVELOPMENT

The process of developing pharmaceutical products is a multivariate optimization problem. It entails optimizing formulation and process variables. The ability of ANNs to generalize is one of their most useful properties. These characteristics make them suitable for resolving formulation optimization problems in pharmaceutical product development. In studies of the effects of various factors (such as formulation and compression parameters) on tablet properties, ANN models demonstrated better fitting and predicting abilities in the development of solid dosage forms (such as dissolution). ANNs proved to be a useful tool for the development of microemulsion-based drug delivery systems with minimal experimental effort (57).

2.10 AI IN THE PHARMACEUTICAL VALUE CHAIN

- *Drug Discovery:* AI applies data science and ML to massive datasets, driving rapid discovery of new molecules. AI in drug discovery cross references published scientific literature and alternative information sources; this includes clinical trial information, conference abstracts, public databases, unpublished data, etc., to innovate potential therapies. Such capabilities have already delivered new drug discoveries at a rapid pace with least turnaround time (58).

- *Clinical Trials:* By automating clinical trials, AI drastically reduces cycle times and costs; it also enhances the outcomes of clinical development. As a result, AI and ML are being adopted by pharma players to automatically generate artifacts. It includes study protocols. AI also leverages NLP to accelerate manual tasks. AI algorithms, when combined with an effective digital infrastructure, effectively enable continuous streams of clinical trial data; this can be cleaned, aggregated, coded, stored, and managed. AI drives faster, smarter, safer, and considerably less expensive clinical trials (59).

- *Operations:* AI optimizes pharma manufacturing facilities and processes. It enables life sciences companies to streamline factory and sensor data to analytics engines which in turn generate new insights. These insights often help enterprises predict process bottlenecks, pinpoint quality control issues, and proactively offer course-correction measures. AI can also better predict demand and supply; it can recommend the next best action to supply chain operators and in certain scenarios autonomously perform specific activities (60).

- *Marketing:* AI gives out smarter, actionable insights to pharmaceutical entities. AI has the power to deliver quality analytics and insights. AI can combine and analyze patient journeys, advertising metrics, and data to optimize omnichannel marketing messaging and channels. In fact, AI facilitates customization of dynamic personalization and engagement of health care professionals (HCPs) based on insights from big data and patterns. AI in certain cases provides recommendations to marketing and sales reps; it may span next best actions, channels, and personalized content (61).

- *Vigilance:* The staggering volume of adverse events impacting the pharma sector is growing by about 10–15% annually; it is compounded by data generated from new sources, like wearables and social media. Processing these adverse event reports often proves to be a manual, costly, and decentralized process. It jeopardizes compliance with safety regulations and impacts early discovery. AI-powered automation coupled with centralization of the intake of adverse event reports can help companies immensely. AI-enabled tools like optical character recognition and NLP drastically reduce case documentation workload; it significantly expedites the investigation process (62).

- *Cyber:* In an era signified by increasing cyber threats, AI has a major role to play. Insider threats are now evading signature-based systems; parallelly, bad actors are using AI to avoid detection by mastering the fundamental detection rules. Smart, AI-driven cyber technologies are needed to protect the big data at the disposal of pharmaceutical and healthcare companies. AI-based approaches can enhance threat intelligence and prediction, thus enabling rapid attack detection and response (63).

2.11 CORRELATION BETWEEN THE FORMULATION FACTORS AND RELEASE PROFILES

- *Beads, Pellets, and Microspheres:* Several ANN models have also been developed to predict the dissolution profiles of controlled-release particulates such as beads, pellets, and microspheres. A CAD/Chem software-created model was used to simulate the effects of processing parameters and process on the release profile of verapamil from multiparticulate beads. The drug release data of the optimized formulations agreed well with those anticipated by the ANN model. An ANN model was used to assess the effect of process parameters on papain entrapment in cross-linked alginate beads, leading to an improvement in stabilization and site-specific delivery. Aspirin-loaded calcium alginate floating microspheres were designed using ANNs and RSM, with formulation material amounts and microsphere release and floating rate used as inputs and outputs, respectively. Compared to RSM, ANNs predicted the *in vitro* drug release profile more accurately. The pH of the external aqueous phase was shown to be a determining factor of incorporation efficiency and drug release behavior in the composition of verapamil hydrochloride-loaded polymer microspheres (64).

- *Solid Dispersions:* The preparation of solid dispersions is a promising method for increasing drug solubility. For the preparation of solid dispersions, a combination of ANNs and a mixture of experimental designs were used. Carbamazepine-Soluplus®-poloxamer 188 solid dispersions were prepared using the solvent casting method to improve the carbamazepine dissolution rate. The influence of solid dispersion composition on carbamazepine dissolution rate was studied using a three-layer feed-forward MLP network as well as a mixture experimental design (65). The mixture experimental design and ANN model accurately described the relationship between the structure of solid dispersion and the percentage of the released drug; however, the MLP network showed excellent predictability than the blend experimental design. A feed-forward back-propagation ANN of logistic sigmoid activation function was used to create a correlation between both the factors and dissolution characteristics as well as optimize dissolution rate for the preparation of PVP/PEG mixtures as carriers for the growth of drug solid dispersions. The prepared solid dispersions demonstrated long-term physical stability, and the ANNs demonstrated adequate prediction power. The effervescent controlled-release floating tablet formulations were created using a nimodipine–PEG solid dispersion.

- *Implants:* Coating of cochlear implants for topical drug delivery systems against inflammation or infectious diseases has been proposed for the management of postsurgical issues associated with cochlear implantation. An ANN model was developed to estimate the formulation parameters and dexamethasone release profile from cochlear implant coatings (66).

- *Liposome:* The use of echogenic liposomes for chemotherapeutic delivery has grown in popularity over the last decade. Since multiple drug resistance can be overcome through controlled drug release, an ANN-based model predictive controller for constant release of the chemotherapeutic agent and maintaining an appropriate concentration at the tumor site has been proposed, which may lessen the occurrence of multidrug resistance and shorten the duration of treatment. ANNs and the MLR method were compared to design and optimize formulation parameters of leuprolide acetate-loaded liposomes (67).

- *Transdermal Formulations:* For optimizing transdermal ketoprofen hydrogel, an ANN framework was designed. For standardizing gel composition, an optimum value of the response variables was used. The optimal formulation's results were in good agreement with those predicted by the ANN model. ANNs have been used to optimize the vehicle composition for transdermal melatonin delivery. The transdermal route was chosen to avoid melatonin's remarkable hepato-gastrointestinal first-pass metabolism and also to maintain steady-state plasma concentrations for acceptable periods. Since melatonin cannot pass through the dense lipophilic matrix of the stratum corneum, several solvents and their mixtures have been used to increase melatonin flux and decrease lag time (68).

- *Microparticles:* To overcome the dissolution rate-limiting step, benznidazole chitosan microparticles were prepared using the coacervation method, followed by the application of an ANN model for formulation optimization. Multi-response optimization was used to achieve the highest yield, encapsulation efficiency, dissolution rate, and smallest size. ANN-predicted optimum values agreed with experimental findings, indicating that ANNs are appropriate for developing optimal benznidazole chitosan microparticles. A flexible drug delivery system for more effectual treatment of infectious diseases has been established using ANNs that can be trained online (69).

- *Nanomaterials:* Analyzing the parameters that influence nanoparticle size, loading efficiency, and cytotoxicity may aid in the development of more efficient drug delivery systems. ANNs have been used to anticipate the physicochemical properties of nanoparticles with significance against such a variety of disorders, analyze complex nonlinear relationships and factors affecting the stability or size of nanoparticles, and design models for trying to identify the relationship between these factors influencing the development of controlled-release drug delivery systems. ANNs can be used for modeling and identification of key parameters that affect the size of nanoparticles in a multidimensional space, in addition to demonstrating input–output relationships. An ANN model for examining the factors that influence nanoparticle size has been developed for the preparation of biodegradable nanoparticles of tri-block poly(lactide)–poly(ethylene glycol)–poly(lactide) (PLA–PEG–PLA) copolymer as a drug carrier. Optimization and modeling have been performed using the spherical central composite design for optimizing the formulation of polymer–lipid hybrid nanoparticles for controlled delivery of verapamil hydrochloride and assessing the influence of formulation factors. ANN models have also been used to build a relationship between both the entrapment efficiency of lipid matrix or polymeric nanoparticles and drug binding energy estimation (70).

2.12 SOME INTERESTING FACTS AND FIGURES REGARDING AI

- In 2021, the AI in the healthcare and pharmaceutical market was worth around US$11 billion globally. It was predicted that the global healthcare AI market would be worth almost US$188 billion by 2030. This would be at a CAGR (compound annual growth rate) of 37% from 2022 to 2030. Figure 2.4 shows a 29.4% compound annual growth rate value for the global AI in the pharmaceutical market from 2022 to 2030. As per a survey conducted in 2021, 19% of respondents stated that AI models have been in production for less than two years in their healthcare organization. A further 18% reported that their organization was evaluating AI use cases. And 26% of the respondents agreed that their organization was not actively considering AI as a business solution (71).
- As per a report by Precedence Research (Table 2.1), AI size in the worldwide pharmaceutical market was valued at US$905.91 million in 2021. It is expected to exceed US$9,241.34 million by 2030 powered by a CAGR of 29.4% from 2022 to 2030.
- As per a recent research report, about 50% of global healthcare companies plan to implement AI strategies and broadly adopt the technology by 2025.
- Recent studies indicate that almost 62% of healthcare organizations are thinking of investing in AI in the near future; and 72% of companies believe AI will be their key business driver in the near future (72).

The small drug molecules sector overshadowed the market with more market share in 2020, compared to large molecules (Figure 2.5).

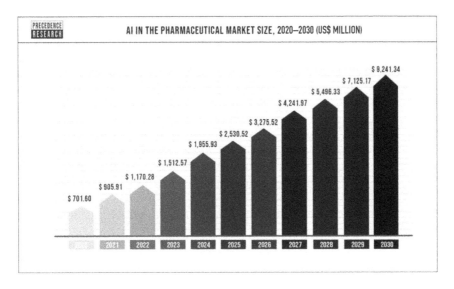

FIGURE 2.4 AI in the pharmaceutical market size, from 2020 to 2030 (US$ million). (https://www.precedenceresearch.com/ai-in-pharmaceutical-market.)

TABLE 2.1

Scope of AI in Pharmaceutical Market as per Precedence Research

Report Coverage	Details
Market Size in 2020	*US$701.6 Million*
Growth rate from 2021 to 2030	29.4%
Revenue projection by 2030	US$9,241.34 million
Largest market	North America
Fastest growing market	Asia Pacific
Base year	2021
Forecast period	2021–2030
Companies mentioned	IBM Corporation, Exscientia, Deep Genomics, Cloud Pharmaceuticals, Inc., Microsoft Corporation, NVIDIA Corporation, Insilico Medicine, Alphabet Inc., Atomwise, Inc., Biosymetrics, Euretos, BenevolentAI

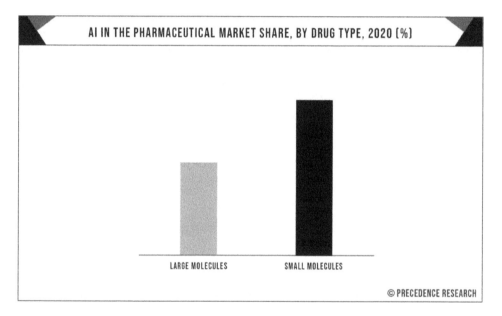

FIGURE 2.5 AI in the pharma market share by drug type: 2020 by percentage. (https://www.precedenceresearch.com/ai-in-pharmaceutical-market.)

2.13 KEY DEVELOPMENTS

- *January 2022:* PostEra, one of the leading biotechnology firms specializing in ML for drug discovery, collaborated with Pfizer to expand its inhouse AI Lab (73).
- *November 2021:* Cyclica and Bayer collaborated to develop an AI-integrated network to optimize drug discovery processes (74).
- *April 2021:* Abbott introduced its new coronary imaging platform, powered by AI in Europe market (75).
- *April 2020:* GSK entered a collaboration with Vir Biotechnology to optimize the process of drug discovery for COVID-19 driven by AI and CRISPR (76).
- *December 2019:* Almirall, a Spanish pharma major, entered a strategic research partnership with Iktos for development of new drug design via AI (77).
- *May 2018:* Pfizer initiated a strategic alliance with XtalPi for AI drug modeling (78).

2.14   AI IN PHARMA MARKET SEGMENTATION (79)

- *By Technology*
  - Context-aware processing
  - Natural language processing
  - Querying method
  - Deep learning
- *By Application*
  - Clinical trial research
  - Drug discovery
  - Research and development
  - Others
- *By Geography*
  - Asia Pacific
    - China
    - India
    - Japan
    - Australia
    - Indonesia
    - South Korea
    - North America
    - The United States
    - South America
    - Brazil
    - Western Europe
    - France
    - Germany
    - UK
    - Eastern Europe
    - Russia
    - Middle East
    - Africa

2.15   OPPORTUNITIES AND RECOMMENDATIONS IN AI PHARMA MARKET (80)

- *Opportunities*
  - AI in pharma technology segment will arise in deep learning, which will gain $721.5 million of worldwide annual sales by 2025 (81).
  - AI in pharma drug-type segment will arise in small molecules, which will gain $1,298.9 million of worldwide annual sales by 2025 (82).
  - AI in pharma application segment will arise in drug discovery, which will gain $1,710.0 million of worldwide annual sales by 2025 (83).
  - AI in overall pharma market size will gain the most in the US region at $665.1 million (84).
- *Recommendations* (85)
  - Focus on investing in AI innovation labs.
  - Use AI technology for developing drugs for rare diseases to reduce the disease burden.
  - Establish operations in emerging markets to gain market share.
  - Scale up through mergers and acquisition.
  - Offer competitive pricing.
  - Increase visibility through websites.
  - Focus on end-user industries.

2.16   AI COMPETITIVE LANDSCAPE (86)

Major competitors in AI in pharmaceutical market are as follows:
- Alphabet, Inc.
- Concreto HealthAI
- IBM Watson Health
- Nvidia Corporation
- PathAI

2.17   OTHER COMPETITORS IN THE AI IN PHARMA MARKET (87)

- Alliance Pharm Pte Ltd.
- Ardigen
- AstraZeneca (South Africa)
- Atomwise, Inc.
- Aurobindo Pharma Limited
- Bayer
- Benevolent AI
- BenevolentAI
- BERG
- Berg LLC
- Bioage
- Botkin
- Care Mentor AI
- China National Pharmaceutical Group Co. Ltd. (Sinopharm)
- Cipla Limited
- Cloud Pharmaceuticals
- Cyclica
- Deep Genomics (Canada)
- Deep Genomics, Inc.
- DeepMind
- Diagnocat
- DilenyTech
- Dr. Reddy's Laboratories Ltd.
- Envisagenics
- Epitrack
- Exscientia
- GE Healthcare
- Gero
- Gesto
- GlaxoSmithKline
- Google
- GSK Australia
- hearX Group
- IBM Corporation
- IBM Watson
- Iktos
- iNNOHEALTH Technology Solutions
- Insilico Medicine
- Intel
- Intensicare

- IQVIA
- Jiangsu Hengrui Medicine Co., Ltd.
- Merck
- Microsoft
- Novartis
- Novartis
- Nucleai
- NuMedii
- Oncora Medical
- Otsuka Pharmaceutical Co. Ltd.
- OWKIN
- Owkin, Inc.
- Pepticom
- Pfizer
- PRA Healthscience
- Recursion Pharmaceuticals Inc.
- Renalytix AI
- Roche
- Roche India
- Rology
- Sanofi
- Shanghai Pharmaceuticals
- Standigm
- Sun Pharma
- Takeda Pharmaceutical Company Ltd.
- Verge Genomics
- Webiomed
- Welltok, Inc.
- XtalPi

2.18 CONCLUSION

Humans have always been said to have hyper intelligent mind power capable of developing innovative technologies that keep adapting to changing times and trends (88). Yet, AI is quickly gaining popularity and attention across diverse domains, especially in healthcare and pharmaceutical industry (89). AI has been described as the fourth industrial revolution framework, with the potential to change the face of every industry (90). AI has evolved as an essential component of the pharmaceutical and broader healthcare domains (91). With numerous ongoing studies on improving the efficiency of manufacturing and other healthcare-related activities, researchers are looking into the possibility of employing AI in multi-interdisciplinary streams to drive productivity, efficiency, and error minimization, thus reducing wastage, ensuring better product quality, and a higher profit margin for companies (92). The COVID-19 pandemic presented numerous opportunities and challenges, including the development of new drugs, drug repurposing, the incorporation of AI in limiting the spread, reducing the cytokine storm, reducing the number and severity of symptoms, and predicting the patient's need for treatment, hospital admission, or ventilation systems (93). AI applications have primarily focused on deep learning–based virtual screening of compounds to improve the scalable synthesis of existing drug treatments (94). The combination of science and data-driven prioritization by AI provides the healthcare system with a helping hand in identifying candidate treatments. As the virus was being pushed away, data limitations hampered the efforts, as data for AI models and projections took months, and public health results also weren't aggregated at the city or regional levels for weeks (95). The efficacy of AI is represented by the combination of technology and the ability of an organization to use it productively and leverage techniques (96). The 2020 pandemic exposed limitations in data workflows, stifling AI progress; but investments are already improving data and analytics infrastructure (97, 98). With increased investments to drive AI throughout the drug discovery and clinical trials, AI has the potential to reshape the core functionality and future of the pharmaceutical industry (99, 100). The use of big data analytics in healthcare is an inevitable reality that necessitates strong regulations, which in turn facilitate data availability and sharing (101).

REFERENCES

1. Chen, Z.; Liu, X.; Hogan, W.; Shenkman, E.; Bian, J. Applications of artificial intelligence in drug development using real-world data. Drug Discov. Today 2020, 26, 1256–1264.
2. Crown, W.H. Real-world evidence, causal inference, and machine learning. Value Health 2019, 22, 587–592.
3. European Medicines Agency. HMA-EMA Joint Big Data Taskforce Phase II Report: "Evolving Data-Driven Regulation."
4. Rivera, S.C.; Liu, X.; Chan, A.-W.; Denniston, A.K.; Calvert, M.J.; Darzi, A.; Holmes, C.; Yau, C.; Moher, D.; Ashrafian, H.; et al. Guidelines for clinical trial protocols for interventions involving artificial intelligence: The SPIRIT-AI extension. Nat. Med. 2020, 26, 1351–1363.
5. Liu, X.; Rivera, S.C.; Moher, D.; Calvert, M.J.; Denniston, A.K.; Chan, A.-W.; Darzi, A.; Holmes, C.; Yau, C.; Ashrafian, H.; et al. Reporting guidelines for clinical trial reports for interventions involving artificial intelligence: The CONSORT-AI extension. Nat. Med. 2020, 26, 1364–1374.
6. Simões, M.F.; Silva, G.; Pinto, A.C.; Fonseca, M.; Silva, N.E.; Pinto, R.M.; Simões, S. Artificial neural networks applied to quality-by-design: From formulation development to clinical outcome. Eur. J. Pharm. Biopharm. 2020, 152, 282–295.
7. Gams, M.; Horvat, M.; Ožek, M.; Luštrek, M.; Gradišek, A. Integrating artificial, and human intelligence into tablet production process. AAPS PharmSciTech. 2014, 15, 1447–1453.
8. Zhang, W.-W.; Li, L.; Li, D.; Liu, J.; Li, X.; Li, W.; Xu, X.; Zhang, M.J.; Chandler, L.A.; Lin, H.; et al. The first approved gene therapy product for cancer Ad-p53 (Gendicine): 12 years in the clinic. Hum. Gene Ther. 2018, 29, 160–179.
9. Obermeyer, Z.; Emanuel, J. Predicting the future: Big data, machine learning, and clinical medicine. N. Engl. J. Med. 2016, 375, 1216–1219.
10. Arsalan, T.; Koshechkin, K.; Lebedev, G. Scientific approaches to the digitalization of drugs assortment monitoring using artificial neural networks. In Czarnowski, I., Howlett, R., Jain, L. (eds) Intelligent Decision Technologies, Smart Innovation, Systems and Technologies; Springer: Singapore, 2020; Volume 193, pp. 391–401.
11. Report Drug Shortages: Root Causes and Potential Solutions, FDA [Electronic Resource]. 1 August 2022.
12. Schork, N.J. Artificial intelligence and personalized medicine. In Precision Medicine in Cancer Therapy; Springer: Cham, Switzerland, 2019; Volume 178, pp. 265–283.
13. Lebedev, G.; Fartushnyi, E.; Fartushnyi, I.; Shaderkin, I.; Klimenko, H.; Kozhin, P.; Koshechkin, K.; Ryabkov, I.; Tarasov, V.; Morozov, E.; et al. Technology of supporting medical decision-making using evidence-based medicine and artificial intelligence. Procedia Comput. Sci. 2020, 176, 1703–1712.
14. Ferrari, D.; Milic, J.; Tonelli, R.; Ghinelli, F.; Meschiari, M.; Volpi, S.; Matteo, F.; Giacomo, F.; Vittorio, I.; Dina, Y.; et al. Machine learning in predicting respiratory failure in patients with COVID-19 pneumonia: Challenges, strengths, and opportunities in a global health emergency. PLoS ONE. 2020, 15, e0239172.
15. Kim, H.-J.; Han, D.; Kim, J.-H.; Kim, D.; Ha, B.; Seog, W.; Lee, Y.-K.; Lim, D.; Hong, S.O.; Park, M.-J.; et al. An easy-to-use machine learning model to predict the prognosis of patients with COVID-19: Retrospective cohort study. J. Med. Internet Res. 2020, 22, e24225.
16. Basile, A.O.; Yahi, A.; Tatonetti, N.P. Artificial intelligence for drug toxicity and safety. Trends Pharmacol. Sci. 2019, 40, 624–635.
17. Schmider, J.; Kumar, K.; Laforest, C.; Swankoski, B.; Naim, K.; Cauble, P.M. Innovation in pharmacovigilance: Use of artificial intelligence in adverse event case processing. Clin. Pharmacol. Ther. 2018, 105, 954–961.
18. Mockute, R.; Desai, S.; Perera, S.; Assuncao, B.; Danysz, K.; Tetarenko, N.; Gaddam, D.; Abatemarco, D.; Widdowson, M.; Beauchamp, S.; et al. Artificial intelligence within pharmacovigilance: A means to identify cognitive services and the framework for their validation. Pharm. Med. 2019, 33, 109–120.
19. Shamseer, L.; Moher, D.; Clarke, M.; Ghersi, D.; Liberati, A.; Petticrew, M.; Shekelle, P.; Stewart, L.A.; PRISMA-P Group. Preferred reporting items for systematic review and meta-analysis protocols (PRISMA-P) 2015: Elaboration and explanation. BMJ. 2015, 350, g7647.

20. Sievert, C.; Shirley, K. LDAvis: A Method for Visualizing and Interpreting Topics; Association for Computational Linguistics. In Proceedings of the Workshop on Interactive Language Learning, Visualization, and Interfaces, Baltimore, Maryland, USA, 27 June 2014; pp. 63–70.
21. Aliper, A.; Plis, S.; Artemov, A.; Ulloa, A.; Mamoshina, P.; Zhavoronkov, A. Deep learning applications for predicting pharmacological properties of drugs and drug repurposing using transcriptomic data. Mol. Pharm. 2016, 13, 2524–2530.
22. Vyasa Analytics Introduces Synapse, A Novel "Smart Table" Software Application Powered by Deep Learning AI Technologies. 1 August 2022.
23. AI-Powered Healthcare. Available online: https://xbird.io/ (accessed 1 August 2022).
24. Xconomy: Gene Network Sciences Using Supercomputing to Match Patients with a Drug that Works. 1 August 2022.
25. Using AI to Accelerate Drug Discovery. Available at: https://www.nature.com/articles/d42473-020-00354-y (accessed 1 August 2022). Appl. Sci. 2022, 12, 8373.
26. Standigm and SK Chemicals Repurpose FDA-approved Drugs into Rheumatoid Arthritis Candidates and Apply for Patents through Their Open Innovation Partnership, Business Wire. Available at: https://www.businesswire.com/news/home/20210107005121/en/Standigm-and-SK-Chemicals-Repurpose-FDA-approved-Drug-into-Rheumatoid-Arthritis-Candidate-and-Apply for-Patent-through-their-Open-Innovation-Partnership (accessed 1 August 2022).
27. Unlearn: Home. Available at: https://www.unlearn.ai (accessed 1 August 2022).
28. Deep Lens BioPharma and CROs. Available at: https://www.deeplens.ai/deep-lens-healthcare-biopharma-cro (accessed 1 August 2022).
29. Feinberg, N.; Sur, D.; Wu, Z.; Husic, B.E.; Mai, H.; Li, Y.; Saisai, S.; Jianyi, Y.; Bharath, R.; Pande, V.S. PotentialNet for molecular property prediction. ACS Cent. Sci. 2018, 4, 1520–1530.
30. Product Pipeline: BioXcel Therapeutics, Inc. (BTAI) Available at: https://www.bioxceltherapeutics.com/ (accessed 1 August 2022).
31. InterVenn Biosciences. Clinical Glycoproteomics + Artificial Intelligence. Available at: https://intervenn.com (accessed 1 August 2022).
32. Transforming Patient Safety: Medaware. Available at: https://www.medaware.com (accessed 1 August 2022).
33. Grønning, N. Data management in a regulatory context. Front. Med. 2017, 4, 114.
34. Turn Patient Insights into Actionirody. Available at: https://irody.com (accessed 1 August 2022).
35. Savana: Transform the Free Text of Your Clinical Records into Big Data. Available at: https://www.savanamed.com (accessed 1 August 2022).
36. LabGenius. Available at: https://labgeni.us (accessed 1 August 2022).
37. InsightRX: Precision Dosing Done Right. Available at: https://www.insight-rx.com (accessed 1 August 2022).
38. Novartis CEO Who Wanted to Bring Tech into Pharma Now Explains Why It's So Hard. Available at: https://www.forbes.com/sites/davidshaywitz/2019/01/16/novartis-ceo-who-wanted-to-bring-tech-into-pharma-now-explainswhy-its-so-hard/ (accessed 1 August 2022).
39. Drug Discovery with an AI-augmented Platform: Cyclica. Available at: https://www.cyclicarx.com (accessed 1 August 2022).
40. Artificial Intelligence Startup Healx Gets $10 Million to Find Cures for Rare Diseases. Healthcare Weekly. Available at: https://healthcareweekly.com/healx-artificial-intelligence-pharma (accessed 1 August 2022).
41. IBM Watson Health. AI Healthcare Solutions. Available at: https://www.ibm.com/watson-health (accessed 1 August 2022).
42. Carvalho, J.P.; Curto, S. Fuzzy Preprocessing of Medical Text Annotations of Intensive Care Units Patients. Available at: https://www.inesc-id.pt/publications/10251/pdf (accessed 1 August 2022).
43. LingPipe Home. Available at: http://www.alias-i.com/lingpipe/ (accessed 1 August 2022).
44. The Stanford Natural Language Processing Group. Available at: https://nlp.stanford.edu/software/lex-parser.shtml (accessed 1 August 2022).
45. Meng, H.; Luk, P.-C.; Xu, K.; Weng, F. GLR parsing with multiple grammars for natural language queries. ACM Trans. Asian Lang. Inf. Process. 2002, 1, 123–144.
46. De Marneffe, M.-C.; Maccartney, B.; Manning, C.D. Generating Typed Dependency Parses from Phrase Structure Parses. In Proceedings of the Fifth International Conference on Language Resources and Evaluation (LREC'06). Genoa, Italy. Available at: https://www.researchgate.net/publication/200044364_Generating_Typed_Dependency_Parses_from_Phrase_Structure_Parses (accessed 1 August 2022).

47. Zhang, X.; Lecun, Y. Text Understanding from Scratch. 2016. Available at: https://arxiv.org/abs/1502.01710 (accessed 1 August 2022).

48. Wang, P.; Qian, Y.; Soong, F.K.; He, L.; Zhao, H. Part-of-speech tagging with bidirectional long short-term memory recurrent neural network. arXiv preprint arXiv:1510.06168. 2015.

49. Neural Network. 2015. Available at: https://arxiv.org/abs/1510.06168 (accessed 1 August 2022).

50. He, J.; Nguyen, D.Q.; Akhondi, S.A.; Druckenbrodt, C.; Thorne, C.; Hoessel, R.; Afzal, Z.; Zhai, Z.; Fang, B.; Yoshikawa, H.; et al. ChEMU 2020: Natural language processing methods are effective for information extraction from chemical patents. Front. Res. Metrics Anal. 2021, 6, 654438.

51. Bouhedjar, K.; Boukelia, A.; Nacereddine, A.K.; Boucheham, A.; Belaidi, A.; Djerourou, A. A natural language processing approach based on embedding deep learning from heterogeneous compounds for quantitative structure–activity relationship modeling. Chem. Biol. Drug Des. 2020, 96, 961–972.

52. Staples, M.; Daniel, K.; Cima, M.J.; Langer, R. Application of micro- and nano-electromechanical devices to drug delivery. Pharm. Res. 2006, 23 (5), 847–863.

53. Tarasov, S.D. Natural Language Generation, Paraphrasing, and Summarization of User Reviews with Recurrent Neural Networks., Available at: http://www.dialog-21.ru/digests/dialog2015/materials/pdf/TarasovDS2.pdf (accessed 1 August 2022).

54. Google AI Blog: Open Sourcing BERT: State-of-the-Art Pre-training for Natural Language Processing. Available at: https://ai.googleblog.com/2018/11/open-sourcing-best-state-of-art-pre.html (accessed 1 August 2022).

55. Better Language Models and Their Implications. Available at: https://openai.com/blog/better-language-models/ (accessed 1 August 2022).

56. Nambiar, A.; Heflin, M.; Liu, S.; Maslov, S.; Hopkins, M.; Ritz, A. Transforming the Language of Life: Transformer Neural Networks for Protein Prediction Tasks. In Proceedings of the 11th ACM International Conference on Bioinformatics, Computational Biology and Health Informatics, BCB 2020, New York, NY, USA, 21–24 September 2020; Volume 5, pp. 1–8. Appl. Sci. 2022, 12, 8373.

57. bioRxiv Preprints Can Now be Submitted Directly to Leading Research Journals. Available at: https://phys.org/news/2016-01-biorxiv-preprints-submitted-journals.html (accessed 1 August 2022).

58. Danger, R.; Segura-Bedmar, I.; Martínez, P.; Rosso, P. A comparison of machine learning techniques for detection of drug target articles. J. Biomed. Inform. 2010, 43, 902–913.

59. Bohr, A.; Memarzadeh, K. The rise of artificial intelligence in healthcare applications. In Artificial Intelligence in Healthcare; Academic Press: Cambridge, MA, USA, 2020; pp. 25–60.

60. Piroozmand, F.; Mohammadipanah, F.; Sajedi, H. Spectrum of deep learning algorithms in drug discovery. Chem. Biol. Drug Des. 2020, 96, 886–901.

61. Lee, J.; Yoon, W.; Kim, S.; Kim, D.; Kim, S.; So, C.H.; Kang, J. BioBERT: A pre-trained biomedical language representation model for biomedical text mining. Bioinformatics. 2020, 36, 1234–1240.

62. Levin, J.M.; Oprea, T.I.; Davidovich, S.; Clozel, T.; Overington, J.P.; Vanhaelen, Q.; Cantor, C.R.; Bischof, E.; Zhavoronkov, A. Artificial intelligence, drug repurposing, and peer review. Nat. Biotechnol. 2020, 38, 1127–1131.

63. Zhou, Y.; Wang, F.; Tang, J.; Nussinov, R.; Cheng, F. Artificial intelligence in COVID-19 drug repurposing. Lancet Digit. Health. 2020, 2, e667–e676.

64. AI-driven Repositioning and Repurposing Summit 2021. Available at: https://pharmaphorum.com/events/ai-drivenrepositioning-and-repurposing-summit-2021/ (accessed 1 August 2022).

65. Mohanty, S.; Rashid, M.H.A.; Mridul, M.; Mohanty, C.; Swayamsiddha, S. Application of artificial intelligence in COVID-19 drug repurposing. Diabetes Metab. Syndr. Clin. Res. Rev. 2020, 14, 1027–1031.

66. Deep Learning Takes on Synthetic Biology. Available at: https://wyss.harvard.edu/news/deep-learning-takes-on-syntheticbiology/ (accessed 1 August 2022).

67. Radivojevíc, T.; Costello, Z.; Workman, K.; Martin, H.G. A machine learning automated recommendation tool for synthetic biology. Nat. Commun. 2020, 11, 1–14.

68. Shah, P.; Kendall, F.; Khozin, S.; Goosen, R.; Hu, J.; Laramie, J.; Ringel, M.; Schork, N. Artificial intelligence and machine learning in clinical development: A translational perspective. NPJ Digit. Med. 2019, 2, 1–5.

69. Overview. The Conference Forum. Available at: https://theconferenceforum.org/conferences/webinar-open-ai-gpt3/overview/ (accessed 1 August 2022).

70. Marius, H. State-of-the-Art Image Classification Algorithm: FixEfficientNet-L2. Towards Data Science. Available at: https://towardsdatascience.com/state-of-the-art-image-classification-algorithm-fixefficientnet-l2-98b93deeb04c (accessed 1 August 2022).

71. Gasimov, H. Cellular Image Classification for Drug Discovery. Intelec AI, Medium. Available at: https://medium.com/intelec-ai/cellular-image-classification-for-drug-discovery-4ef55741151c (accessed 1 August 2022).

72. Lebedev, G.; Fartushniy, E.; Shaderkin, I.; Klimenko, H.; Kozhin, P.; Koshechkin, K.; Ryabkov, I.; Tarasov, V.; Morozov, E.; Fomina, I.; et al. Creation of a medical decision support system using evidence-based medicine. Int. Conf. Intell. Decis. Technol. 2020, 413–427.

73. Flasinski M. Introduction to Artificial Intelligence. 1st ed. Switzerland: Springer International Publishing; 2016. p.4.

74. The Beginner's Guide to Artificial Intelligence. 2019, September 13.

75. https://www.chapter247.com/blog/artificial-intelligence-beginners-guide/ (accessed 23 November 2019).

76. Shapiro SC. Artificial intelligence. Encyclopedia of Artificial Intelligence, Vol. 1, 2nd edn. New York; Wiley, 1992.

77. Rachel Brazil. Artificial intelligence: Will it change the way drugs are discovered? Pharmaceut. J. 2017, 299(7908). doi: 10.1211/PJ.2017.20204085

78. Topol, E.J. High-performance medicine: The convergence of human and artificial intelligence. Nat Med. 2019, 25, 44–56. doi: 10.1038/s41591-018-0300-7.

79. Silver, D.; Schrittwieser, J.; Simonyan K. Mastering the game of Go without human knowledge. Nature. 2017, 550, 354–359.

80. Moosavi-Dezfooli, S-M.; Fawzi, A.; Fawzi, O.; Frossard P. Universal adversarial perturbations. IEEE Conference on Computer Vision and Pattern Recognition, Hawaii, 21–26 July 2017.

81. Manikiran, S.S.; Prasanthi N.L. Artificial intelligence: Milestones and role in pharma and healthcare sector. Pharma Times. 2019, 51(1), 9–15.

82. Chethan Kumar G.N. Artificial Intelligence: Definition, Types, Examples, Technologies. Available at: https://medium.com/@chethankumargn/artificial-intelligence-definition-types-examples-technologies-962ea75c7b9b (accessed 23 November 2019).

83. Sutariyaa, V.; Grosheva, A.; Sadanab, P.; Bhatia, D.; Pathaka Y. Artificial neural network in drug delivery and pharmaceutical research. Open Bioinform J. 2013, 7(Suppl-1, M5) 49–62.

84. Agatonovic-Kustrin, S.; Beresford R. Basic concepts of artificial neural network (ANN) modeling and its application in pharmaceutical research. J. Pharm. Biomed. Anal. 2000, 22, 717–727.

85. Baxt, W.G.; Skora J. Prospective validation of artificial neural network trained to identify acute myocardial infarction. Lancet. 1996, 347, 12–15.

86. Kalis, B.; Collier, M.; Fu R. 10 Promising AI Applications in Health Care. Available at: https://hbr.org/2018/05/10-promising-ai-applications-in-health-care (accessed November 23, 2019).

87. Nic Fleming. How artificial intelligence is changing drug discovery. Nature. 2018, 557, S55–S57. doi: 10.1038/d41586-018-05267-x

88. https://www.digitalauthority.me/resources/artificial-intelligence-pharma/ (accessed 23 November 2019).

89. https://www.himssanalytics.org/sites/himssanalytics/files/2017_Essentials%20Brief_Mobile_SNAPSHOT%20REPORT.pdf (accessed 23 November 2019).

90. Keshavan M. Berg: Using Artificial Intelligence for Drug Discovery. http://www.medcitynews.com/2015/07/berg-artificialintelligence/ (accessed 23 November 2019).

91. Cattell, J.; Chilukuri, S.; Levy M. How big data can revolutionize pharmaceutical R and D. Available at: https://www.mckinsey.com/industries/pharmaceuticals-and-medical-products/our-insights/how-big-datacan-revolutionize-pharmaceutical-r-and-d (accessed 23 November 2019).

92. Ozerov, I.V.; Lezhnina, K.V.; Izumchenko E.; et al. In silico Pathway Activation Network Decomposition Analysis (iPANDA) as a method for biomarker development. Nature Commun. 2016, 7, 13427.

93. Bajorath, J.; Kearnes, S.; Walters, W.P.; Georg, G.I.; Wang S. The future is now: Artificial intelligence in drug discovery. J. Med. Chem. 2019, 62 (11), 5249. doi: 10.1021/acs.jmedchem.9b00805

94. Jiang, F.; Jiang, Y.; Zhi, H.; Dong, Y.; Ma, S.; Wang, Y.; Dong, Q.; Shen, H.; Wang Y. Artificial intelligence in healthcare: Past, present, and future. Stroke Vasc. Neurol. 2017, 2, e000101.

95. AI in Pharmacy: Speeding up Drug Discovery. Available at: https://medium.com/sciforce/ai-in-pharmacy-speedingup-drug-discovery-c7ca252c51bc (accessed 23 November 2019).

96. Shahriar, A.; Hossain, M.A.; Sajib, S.; Sultana, S.; Rahman M.; Vrontis, D.; McCarthy, G. A framework for AI-powered service innovation capability: Review and agenda for future research. *Technovation*. 2023, 125, 102768. https://doi.org/10.1016/j.technovation.2023.102768.

97. Jon G. Will Amazon and Robots Take Jobs Away from Pharmacists? Available at: Available at: http://www.hcrnetwork.com/amazon-robots-take-jobs-away-pharmacists/ (accessed 23 November 23, 2019).

98. Modern Marketing: Pharma's Data-powered AI Revolution. Available at: https://uploads-ssl.webflow.com/5aa82d18dd47c40638d41d39/5ba26eea28a3c1298fb91a39_Modern-Pharma-Marketing-Ebook-PDF-Download-9.19-compressed.pdf (accessed 23 November 2019).

99. IDC Study. Data Age 2025. Sponsored by Seagate. April 2017.

100. Khedkar, P.; Mitra S. Boosting Pharmaceutical Sales and Marketing with Artificial Intelligence. Available at: https://www.zs.com/Publications/Articles/Boosting-Pharmaceutical-Sales-and-Marketing-With-Artificial-Intelligence (accessed 23 November 2019).
101. Real-World Applications of Artificial Intelligence to Improve Medication Management across the Care Continuum. Electronic Health Reporter. Available at: https://electronichealthreporter.com/real-world-applications-of-artificial-intelligence-toimprove-medication-management-across-the-care-continuum/.

3 Artificial Intelligence in Pharmaceutical Technology
Scope for the Future

Hanan Fahad Alharbi
Princess Nourah bint Abdul Rahman University, Riyadh, Saudi Arabia

Mullaicharam Bhupathyraaj
National University of Science and Technology, Muscat, Oman

Leena Chacko
Meso Scale Diagnostics LLC, Rockville, Maryland, USA

Kiruba Mohandoss
Sri Ramachandra Institute of Higher Education and
Research, Chennai, Tamil Nadu, India

K. Reeta Vijaya Rani and S Anbazhagan
Surya School of Pharmacy, Vikravandi, Tamil Nadu, India

3.1 BACKGROUND

Artificial Intelligence (AI) is a subfield of computer science that deals with problem-solving using symbolic programming. It has managed to evolve into a problem-solving science with widespread application in business, healthcare, and engineering. AI's main goal is to identify useful information-processing problems and provide an abstract account of how to solve them (1). This type of account is known as a method, and it relates to a hypothesis in mathematics. AI is a field that deals with the design and implementation of algorithms for data analysis, learning, and interpretation. AI covers a wide range of statistical and machine learning, pattern recognition, clustering, as well as similarity-based methods (2).

AI is a burgeoning technology that has applications in a wide range of areas of life and industry. In recent years, the pharmaceutical industry has discovered a novel and innovative way to use this powerful technology to help solve some of the industry's most pressing problems (3). In the pharmaceutical industry, AI refers to the use of automated algorithms to perform tasks that have traditionally relied on human intelligence. In the last five years, the use of AI in the pharmaceutical and biotech industries has transformed how scientists develop new drugs to combat disease and more (4).

Over the last few decades, there has been an interest in the development of innovative systems for the targeted delivery of therapeutics with maximum efficiency and minimal side effects. Controlled drug delivery and tackling the challenges associated with traditional drug delivery systems, such as systemic toxicity, narrow therapeutic index, and dose adjustment in long-term therapy, have been the subject of intensive research (5). The use of microfabrication technology to create implantable microchips holds promise for controlled drug delivery.

DOI: 10.1201/9781003343981-3

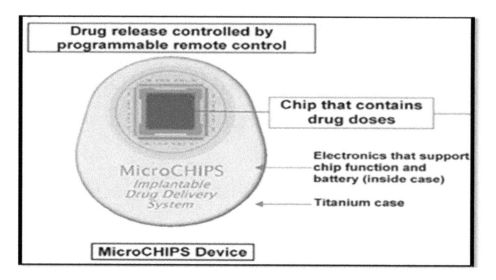

FIGURE 3.1   A microchip-based device for electronic format (8, 9).

By utilizing technological advancements, the pharmaceutical industry can accelerate its development. The most recent technological advancement that comes to mind is AI, which is the development of computer systems capable of performing tasks normally requiring human intelligence, such as visual processing, speech recognition, decision-making, and language translation (6). According to IBM, the entire healthcare domain has 161 billion GB of data as of 2011. With so much data available in this domain, AI can be of great assistance in scrutinizing it and presenting results that will aid in decision-making, saving human effort, time, and money, and thus helping to save lives (7).

The microchip (Figure 3.1) is housed in a biocompatible case that also houses the electronic components, transmitter, and energy source for wireless technology.

3.2   APPLICATION OF AI IN DRUG DELIVERY SYSTEMS

Creating a commercial product, whether that is a relatively simple formulation (e.g., a capsule, a tablet, or an oral liquid) or a controlled-release formulation (e.g., an implant), is always a time-consuming and complex issue (10). Overall, an initial formulation of one or more drugs mixed with various ingredients (excipients) is prepared, and as development progresses, the selection of these and their levels, or even the manufacturing process, is started to change and configured as a result of significant, time-consuming experimental work. These iterations generate a large amount of data, which is difficult to process and understand (11).

In reality, the formulator must work in a multidimensional design space that is nearly impossible to conceptualize. Statistics have been used as one approach to the situation to date. This approach has the advantage of producing models that are clearly expressed and have associated confidence factors (12). However, statistical approaches quickly become unwieldy for more than three or four inputs, so the formulator is tempted to oversimplify the problem (for example, limiting a study to three input variables) to characterize it.

Neural networks are mathematical constructs that can "learn" relationships within data without the user's prior knowledge. The neural network makes no presumptions about the conceptual model of the relationships; instead, it generates and evaluates a variety of models to find the one that best fits the experimental data provided to it (13). As a result, artificial neural networks (ANNs) are increasingly being used to model complex behavior in problems such as pharmaceutical formulation and processing.

Neural networks, like humans, learn directly from input data. There are two types of learning algorithms. Unsupervised learning, in which the network is given input data and learns to recognize patterns in the data, helps organize large amounts of data into fewer clusters. The network is presented with a series of matching input and output examples for supervised learning, which is analogous to "teaching" the network, and it understands the relationships linking the inputs to the outputs (14). Supervised learning has proven particularly useful in formulation, where the goal is to establish cause-and-effect relationships between inputs (ingredients and processing conditions) and outputs (measured properties).

3.2.1   Formulation

- *Controlled-release Tablets:* Hussain and colleagues at the University of Cincinnati pioneered the use of neural networks for modeling pharmaceutical formulations (Ohio, USA). They modeled the *in vitro* release characteristics of a variety of drugs distributed in matrices formulated from varied hydrophilic polymers in several studies. In all cases, neural networks with a single hidden layer were found to perform reasonably well in drug release prediction. Personnel from the pharma industry KRKA d.d. (Smerjeska, Slovenia) and the University of Ljubljana (Slovenia) used neural networks to predict the release of a drug and also to optimize using two- and three-dimensional response surface analyses in a much more recent survey involving the formulation of diclofenac sodium from a matrix tablet formulated from cetyl alcohol (15).
- *Immediate-release Tablets:* Two studies were conducted in this area just three years ago. Turkoglu and colleagues from the University of Marmara (Turkey) and the University of Cincinnati used neural networks and statistics to model hydrochlorothiazide tablet formulations. The networks created were used to generate three-dimensional plots of massing time, compression pressure, and crushing strength, or drug release, to maximize tablet strength or choose the appropriate lubricant (16).

Even though trends were observed, no optimal formulations were provided. The trends resembled those produced by statistical procedures. Genetic algorithms were used to generate comparable neural network models, which were then optimized. The optimum formulation was discovered to be dependent on the constraints applied to the ingredient levels used in the formulation as well as the relative importance placed on the output parameters. Disintegration time was sacrificed to achieve high tablet strength and low friability. Lactose was the recommended diluent in all cases, and a fluidized bed was the preferred granulating technology (16).

3.2.2   Product Development

The process of developing pharmaceutical products is a multivariate optimization problem. It entails optimizing formulation and process variables. The ability of ANNs to generalize is one of their most useful properties. These characteristics make them suitable for resolving formulation optimization problems in pharmaceutical product development. In studies of the effects of various factors (such as formulation and compression parameters) on tablet properties, ANN models demonstrated better fitting and predicting abilities in the development of solid dosage forms (such as dissolution). ANNs proved to be a useful tool for the development of microemulsion-based drug delivery systems with minimal experimental effort (17).

3.3   CONCLUSION

Humans have always been said to have much more intelligent brain power capable of developing innovative technologies that are better and explaining them is a major task. AI is quickly gaining popularity and attention in the healthcare system. AI has been described as the fourth industrial revolution machine, with the potential to change the face of every industry. AI has become an essential component of the pharmaceutical industry as well as the healthcare team (18).

With numerous studies on improving the efficiency of manufacturing and other healthcare-related activities, researchers are looking into the possibility of employing AI in multidisciplinary fields due to productivity, efficiency, and error minimization, thus reducing wastage, better product quality, and a higher profit margin for companies (19).

The COVID-19 pandemic presented numerous opportunities and challenges, including the development of new drugs, the incorporation of AI in limiting the spread, reducing the cytokine storm, reducing the number and severity of symptoms, and predicting the patient's need for treatment, hospital admission, or ventilation systems (20). AI applications have primarily focused on deep learning–based virtual screening of compounds to improve the scalable synthesis of existing drug treatments. The combination of science and data-driven prioritization by AI provides the healthcare system with a helping hand in identifying candidate treatments. As the virus was being pushed away, data limitations hampered the efforts, as data for AI models and projections took months, and public health results also weren't aggregated at the city or regional levels for weeks (21).

The efficacy of AI is represented by the combination of technology and the ability of an organization to use it productively and leverage techniques. The 2020 pandemic exposed limitations in data workflows, stifling AI progress, but investments are already improving data and analytics infrastructure. With increased investments to drive AI throughout the drug discovery and clinical trials, AI has the potential to reshape the pharmaceutical industry. The use of big data analytics in healthcare is an unavoidable reality that necessitates strong regulations that facilitate data availability and sharing (22).

REFERENCES

1. J.F. Dastha, Application of artificial intelligence to pharmacy and medicine, Hosp. Pharm. 27 (1992) 312–315, 319–322.
2. W. Duch, K. Swaminathan, J. Meller, Artificial intelligence approaches for rational drug design and discovery, Curr. Pharm. Des. 13 (2007) 1497–1508.
3. D. Jimenez, How technology could transform drug research in 2022, Pharm. Technol. 12 (2021)1–2. https://www.pharmaceutical-technology.com/features/how-technology-could-transform-drug-research-in-2022/.
4. P. Tangri, S. Khurana, Pulsatile drug delivery systems: Methods and advances, Int. J. Drug Formul. Res. 2 (2011) 100–111.
5. M. Alessandra, D.D. Maria, L. Giulia, Oral pulsatile delivery: Rationale and chronopharmaceutical formulations, Int. J. Pharm. 398 (2010) 1–8.
6. M. Staples, K. Daniel, M.J. Cima, R. Langer, Application of micro- and nano-electromechanical devices to drug delivery, Pharm. Res. 23 (2006) 847–863.
7. K.B. Sutradhar, C.D. Sumi, Implantable microchip: The futuristic controlled drug delivery system, Drug Deliv. 23 (2016) 1–11.
8. J.H. Prescott, S. Lipka, S. Baldwin, N.F. Sheppard Jr., J.M. Maloney, J. Coppeta, B. Yomtov, M.A. Staples, J.T. Santini Jr., Chronic, programmed polypeptide delivery from an implanted, multi-reservoir microchip device, Nat. Biotechnol. 24 (2006) 437–438.
9. R.C. Rowe, E.A. Colbourn, Artificial intelligence the key to process understanding, Pharm. Tech. Eur. 9 (1996) 46–55.
10. M. Melanie, An introduction to genetic algorithms. MIT Press, Cambridge, MA, 1999.
11. C. Hayes, T. Gedeon, Hyperbolicity of the fixed point set for the simple genetic algorithm, Theor. Comput. Sci. 411 (2010) 24–29.
12. R.C. Chakraborty, "Fundamentals of genetic algorithms." AI course 2010, lecture 39–40.
13. D. Goldberg, Genetic algorithms in search, optimization and machine learning. Addison Wesley, 1989.
14. K.F. Man, K.S. Tang, S. Kwong, Genetic algorithms: Concepts and designs, Chapters 1–10. Springer, London, 1999.
15. S. Vaithiyalingam, M.A. Khan, Optimization and characterization of controlled release multi-particulate beads formulated with a customized cellulose acetate butyrate dispersion, Int. J. Pharm. 234 (2002) 179–193.

16. M.G. Sankalia, R.C. Mashru, J.M. Sankalia, V.B. Sutariya, Papain entrapment in alginate beads for stability improvement and site-specific delivery: Physicochemical characterization and factorial optimization using neural network modeling, AAPS PharmSciTech 6 (2005) E209–E222.

17. H.I. Labouta, L.K. El-khordagui, A.M. Molokhia, G.M. Ghaly, Multivariate modeling of encapsulation and release of an ionizable drug from polymer microspheres, J. Pharm. Sci. 98 (2009) 4603–4615.

18. P. Hassanzadeh, F. Atyabi, R. Dinarvand, Application of modeling and nanotechnology-based approaches: The emergence of breakthroughs in theranostics of central nervous system disorders, Life Sci. 182 (2017) 93–103.

19. Y. Sun, Y. Peng, Y. Chen, A.J. Shukla, Adv. Drug Deliv. Rev. 55 (2003) 1201–1215.

20. H. Asadi, K. Rostamizadeh, D. Salari, M. Hamidi, Preparation of biodegradable nanoparticles of triblock PLA–PEG–PLA copolymer and determination of factors controlling the particle size using artificial neural network, J. Microencapsul. 28 (5) (2011) 406–416.

21. Y. Li, M.R. Abbaspour, P.V. Grootendorst, A.M. Rauth, X.Y. Wu, Optimization of controlled release nanoparticle formulation of verapamil hydrochloride using artificial neural networks with genetic algorithm and response surface methodology, Eur. J. Pharm. Biopharm. 94 (2015) 170–179.

22. A.A. Metwally, R.M. Hathout, Computer-assisted drug formulation design: Novel approach in drug delivery, Mol. Pharm. 12 (2015) 2800–2810.

4 Applications of Artificial Intelligence in Drug Delivery Systems

Vijaya Rajendran, Carolin Lincy B.J., and Gothanda RamanG
Anna University, Chennai, Tamil Nadu, India

Kirubakaran Narayanan
SRM College of Pharmacy, Kattankulathur, Tamil Nadu, India

Leena Chacko
Meso Scale Diagnostics LLC, Rockville, Maryland, USA

Mullaicharam Bhupathyraaj
National University of Science and Technology,
Muscat, Oman

Kiruba Mohandoss
Sri Ramachandra Institute of Higher Education and
Research, Chennai, Tamil Nadu, India

Hanan Fahad Alharbi
Princess Nourah bint Abdul Rahman University,
Riyadh, Saudi Arabia

4.1 INTRODUCTION

Over the past few years, there has been a drastic increase in data digitalization in the pharmaceutical sector. However, this digitalization comes with the challenge of acquiring, scrutinizing, and applying that knowledge to solve complex problems, especially in drug delivery systems. Artificial Intelligence (AI) is a technology-based system involving various advanced tools and networks that can help resolve the issues in drug delivery and enhance the therapeutic benefits in the pharmaceutical segments (1). AI utilizes systems and software that can interpret and learn from the input data to make independent decisions for accomplishing specific criteria (2). This includes design of new treatment hypotheses and related strategies, disease progression status, and determining pharmacological actions of drug moieties, drug design, optimizing preformulation of various dosage forms, modeling IVIVC correlations, interpreting analytical data, and determining the relationship between the chemical structure and biological activity of a substance (3). Artificial neural networks (ANNs) involve various subtypes, like multilayer perceptron (MLP) networks, recurrent neural networks (RNNs), and convolutional neural networks (CNNs). ANN models help immensely in recognizing the capabilities of the neural networks of the brain. Similar to a single neuron in the brain, an artificial neuron unit receives inputs from many external sources, processes them, and makes

DOI: 10.1201/9781003343981-4

decisions (4, 5). In this chapter, the applications of AI in drug delivery systems are to be elaborated to understand the critical role of AI in the pharmaceutical field. The various AI networks and tools are as follows:

- *Machine Learning (ML) (6):* The fundamental paradigm of ML involves reasoning, knowledge representation, solution search, etc. The principle behind ML is algorithms.
- *Artificial Neural Networks (7):* This comprises a set of interconnected sophisticated computing elements involving "perceptions" like human biological neurons, which will mimic the transmission of electrical impulses in the human brain.
- *Recurrent Neural Networks (8):* RNNs are networks having the capability to memorize and store information with a closed loop.
- *Convolution Neural Networks (9):* It is a series of dynamic systems characterized by its topology and has been used in image and video processing, biological system modeling, complex brain functions processing, pattern recognition, and sophisticated signal processing with local connections mode.
- *Kohonen Networks (10):* The principle behind the Kohonen network is to map the input of patterns of arbitrary dimension N onto a discrete map with one or more dimensions. The applied patterns close to one in the input space should be close to one another in the application map: they should be topologically ordered. A Kohonen network contains a grid of output units and N input units. The input pattern is fed to each output unit. The input lines to each output unit are calculated based on the weight. These weights are initialized into small random numbers for application purposes in AI segments. The application of Kohonen neural networks (KNN) in pharmaceuticals is to classify biologically interesting compounds in drug delivery. The database includes more than 2,000 organophosphorus-potent pesticides (11). The Kohonen maps help to distinguish biological compounds with different pharmacological actions.
- *Radial Basis Function (RBF) Networks:* RBF network in its simplest form is a three-layer feed-forward neural network (12). The first layer corresponds to the inputs of the network, the second is a hidden layer consisting of a number of RBF nonlinear activation units, and the last one corresponds to the final output of the network. A control strategy using the RBF network has been the main application in an Ibuprofen production process. This helps to understand the laboratory model in mini scale or pilot crystallization aspect of Ibuprofen in the pharmaceutical process (13).
- *Learning Vector Quantization (LVQ) Networks:* LVQ is one of the ANNs which is also inspired by biological models of neural systems. The principle behind this tool is to prototype a supervised learning classification algorithm and train its network through a competitive learning algorithm. It helps resolve the multiclass classification problems. LVQ has worked on two layers: one layer is the input and the other one is the output (14).
- *Counter-propagation Networks (CPN):* CPN works on a multilayer network principle based on the combinations of the input, output, and clustering layers. The application of CPN spans data compression, function approximation, and pattern association. Drug-induced liver injury is a major concern in the drug development process. Expensive and time-consuming *in vitro* and *in vivo* studies do not reflect the complexity of the phenomenon (15).

    The capabilities of counter-propagation artificial neural networks (CPANNs) for the classification of an imbalanced dataset related to idiosyncratic drug-induced liver injury is vital. It is also useful to develop a model for the prediction of the hepatotoxic potential of drugs. Genetic algorithm optimization of CPANN models was used to build models for the classification of drugs into hepatotoxic classes and non-hepatotoxic classes using molecular descriptors. In addition, counter-propagation ANN categorical models were used for prediction of carcinogenicity for non-congeneric chemicals (16).

- *Adaptive Linear Neuron or Later Adaptive Linear Element (ADALINE) Network:* ADALINE is an early single-layer ANN. The architecture incorporated into this machine consists of a single neuron with some inputs connected to it. A multilayer network of ADALINE units is known as a MADALINE (17).
- Multilayer perception (MLP): MLP application facilitates pattern recognition, optimization aids, and process identification and control aspects in pharmaceutical drug delivery.

4.1.1 Prediction of the Target Protein Structure

During design and delivery of a drug moiety, it is essential to assign the exact target for successful treatment schedule. More number of proteins which involve in the development of the disease are stated; and in some cases, they are overexpressed conditions. Hence, to design the drug for targeting of disease profile, it is more important to predict the structure of the target protein. AI can help understand the structure-based drug discovery by predetermining the 3D protein structure. One of the AI tools, named Alpha Fold, is based on DNN principle, and this application is used to determine the distance between the adjacent amino acids as well as the corresponding angles of the peptide bonds. It is to ensure the presence of 3D target protein structures in drug delivery. AlQurashi studied the RNN-based tools for the protein structure. Computation, geometry, and assessment were found to be some of the recurrent elements in geometric network (RGN). As part of this protein model study, the primary protein sequence was encoded. The torsional angles for a given residue as well as a partially completed backbone (obtained from this geometric unit upstream) were input and provided a new backbone as output. The final 3D structure unit was the output. Assessment of the deviation of predicted and result structures was done using the distance-based root mean square deviation (dRMSD) metric as illustrated in Figure 4.1 (18–21).

4.1.1.1 Predicting Drug–Protein Interactions Using AI

Drug–protein interactions are particularly important to predict the drug pharmacological pathway. The prediction of drug interaction with a receptor or a protein is essential to understand its efficacy and the repurposing of drugs for the prevention of polypharmacology (22). A study was done by Wang et al. using the SVM model, trained on 15,000 protein–ligand interactions; it was developed based on primary protein sequences and structural characteristics of small molecules to discover new compounds as well as their interaction with target proteins (23).

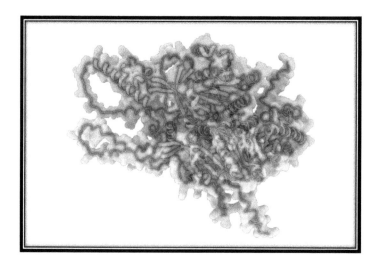

FIGURE 4.1   Protein structure. *Source:* https://www.science.org/content/article/ai-cracks-code-protein-complexes-providing-road-map-new-drug-targets.

4.2 CELLULAR NETWORK-BASED DEEP LEARNING (DL) TECHNOLOGY (deepDTnet)

This technique was used to predict the therapeutic use of topotecan, which is used as a topoisomerase inhibitor. It can also be used for the therapy of multiple sclerosis by inhibiting human retinoic acid receptor–related orphan receptor-gamma t (ROR-gt) (24).

4.3 SELF-ORGANIZING MAPS (SOMs)

It is an unsupervised category of ML and the main application in pharmaceuticals is drug repurposing. The underlying principle is a ligand-based approach to search novel off-targets for a set of drug molecules by training the system on a defined number of compounds with recognized biological activities. It is used for the analysis of different compounds (25–27).

4.4 AI-INBUILT NANOROBOTS FOR DRUG DELIVERY

These are integrated circuits, sensors, power supply, etc. that secure backup of data, which are maintained via computational technologies called nanorobots. It is programmed to avoid a collision, target identification, detect and attach, and finally excrete from the body. Advances in nano/micro-robots have enabled to navigate to the targeted site which is based on physiological conditions like pH, thus improving the efficacy and reducing systemic adverse effects. Application of implantable nanorobots developed for controlled delivery of drugs and genes requires consideration of parameters, mainly dose adjustment, sustained release, control release, and the release of the drugs which require automation (NNs, fuzzy logic, integrators, etc.) (28, 29).

4.5 AI IN COMBINATION DRUG DELIVERY AND SYNERGISM/ANTAGONISM PREDICTION

Some of the combinations of drugs are approved by licensing authorities and marketed to treat complex disease conditions like cancer (30), HIV infections, and tuberculosis, due to the synergistic effect for immediate recovery and long-time treatment therapy. Some AI tools like ANNs, logistic regression, and network-based modeling help to screen drug combinations as well as improve the overall dose regimen for long-term therapy. A case study published by Rashid et al. was related to quadratic phenotype optimization platform to determine optimal combination therapy for the treatment of bortezomib-resistant multiple myeloma. This technique of tools was used to identify the best suitable combination of decitabine (Dec) and mitomycin C (MitoC) as the most suitable drug combinations among the existing 114 FDA drugs.

4.6 AI EMERGENCE IN NANOMEDICINE

The principle behind the nanomedicines is delivery of medicines through nanotechnology for the applications of diagnosis, treatment, and monitoring of complex disease conditions like cancer, HIV, asthma, and malaria. Currently, in the pharmaceutical segment, nanoparticle-modified drug delivery has become important in the areas of therapeutics and diagnostics as they have enhanced efficacy and treatment beneficiaries for chronic disease conditions. A combined technique of nanotechnology which is assisted by AI could provide solutions to many challenges in design and development of drug formulations (31–33).

4.6.1 BLOOD–BRAIN BARRIER (BBB) PERMEABILITY

The delivery of drug molecules through the BBB can be determined by a computational technique using 75–97% accuracy (for penetrating and nonpenetrating molecules) using 67–199 data descriptors (34). According to one of the case studies of Zhao et al., accuracy for the nonpenetrating molecules is smaller (60–80%); this bias in the statistical learning methods was resolved by using recursive feature elimination to choose the features through only 19 molecular descriptors like polarizability, polarity-related properties, hydrogen bond properties, volume, weight, surface area, bond rotations, and pK_a. The molecules in their training set were classified with an accuracy that exceeded 90%. Their model predicted the penetration of the drug molecules with BBB in a test set with an accuracy that exceeded 95%. In another study, Garg and Verma built a MLP using the ANN technique using seven descriptors (35). The molecular weight and the topology of the polar surface area were some of the important descriptors and predicted, on a test set, the BBB ratios with a correlation coefficient of 0.89 as a static output.

4.7   AI AND ML APPLICATIONS IN INTRACTABLE DISEASES

- *COVID-19:* In current pandemic conditions to diagnose COVID-19 cases, AI models have been designed successfully. In one of the case studies, a neural network was designed to determine, from CT scans combined using clinical data, whether or not a patient tested positive for COVID-19. This neural network has two separate model combinations using deep CNN to detect certain peculiarities in the CT scans, as well as SVM, MLP, and random forest in the segmentations of the patients according to the collected data in clinical trials (36). In another case study, Gao et al. revealed that 96.1% similarity of drugs reported in chemotherapies applicable to SARS-CoV should be useful in treating SARS-CoV-2. In addition, network complex ML models were leveraged to generate new drugs for COVID-19 using already existing 115 SARS-CoV protease inhibitors from the ChEMBL database by repurposing pharmacology principles (37).
- *Cancer:* One of the case studies of the European INSPIRE project focuses on personalized drug therapy for precise cancer treatment regimens using ML application tools. ML is a key tool agent in the pharmaceutical segment, especially in drug design, for personalized treatment of cancer disease conditions. This is due to 10,000–100,000 mutations that can be simulated to maximize efficiency on an individualized basis (38). The correlation between the disease and the gene is a useful tool to address cancer in the segment of "cancer genomics." In cancer genomics, DL with multiple layers is advantageous as it can handle the complexity of cancer-related studies.

4.7.1   Optimizing Drug Delivery Using AI

Drug delivery has various techniques like formulations, manufacturing techniques, storage of bioactive compounds, and transportation to the target sites, to achieve a maximum therapeutic benefit (39). Currently, drug delivery strongly depends on nanotechnology-based innovations. Scientists have been applying various knowledge of sciences, like pharmaceutical sciences, biosciences, and engineering, to optimize drug delivery for desired actions with the additional support of computational methods (40–42). AI application is now being used for analyzing and interpreting biological and genetic information to deliver drugs (43–51).

4.8   ANALYTICAL DATA ANALYSIS AND STRUCTURE RETENTION RELATIONSHIP (SRR)

ANNs are an important AI tool to recognize patterns from complex analytical data; they use data analysis in the segment of pharmacological research, especially nonlinear relationships from noisy data. In addition, ANNs help in spectral data of the multicomponent sample analysis process. Some of the applications of ANNs include identification and calibration of amino acids with the same structured molecule and spectrums with the example of tryptophan and tyrosine (52–54). ANNs are useful in chiral sample analysis and enantiomeric determinations using the principle of nonlinear relationships. ANNs help in the method of quantification of antihistamines like ranitidine hydrochloride by the mechanism of IR spectrum and X-ray diffraction technique with different forms of ranitidine (55). ANN helps in facilitating the examination of Sequest's results of MS/MS data for peptide studies. For anion separations in chromatography, computer-assisted optimization method of ANNs is used to select the optimal gradient determinations. ANNs application is also useful in rapid and accurate prediction of retention times for anions in ion-exchange chromatography. ANNs can be a useful tool in HPLC for the optimization of chromatographic conditions like capacity factors, mobile-phase composition, and pH to determine the retention times. ANNs are also useful in the detection of two similar bisphenols in blood level study using generic algorithms (GA) tools (56). ANNs are extensively used in the optimization stage of the method development of pharmaceutical formulations and characterization of pharmaceutical mixtures.

4.8.1 PREFORMULATION

ANN tools are used in designing the optimization studies and preformulation studies of amorphous polymers (57). ANN helps in the prediction of relationships between the compositions of polymer blend and the water uptake profiles in the selection of active pharmaceutical ingredients (API). ANN is also used in the study of fast-release tablet databases and the effect of API on tablet dosage forms (58, 59). ANNs are also used in the design of stable formulations in multiple active components within emulsion dosage form.

4.8.2 OPTIMIZATION OF PHARMACEUTICAL FORMULATIONS

The prediction of pharmaceutical responses based on the polynomial equation as well as response surface methodology (RSM) has been widely used in formulation optimization in pharmaceutical industry (60–64). ANNs have been widely used to optimize controlled-release capsules and predict the release profile in aspirin extended-release tablets.

4.8.3 IVIVC

IVIVC will play an important role in the pharmaceutical segment to avoid bioequivalence studies. Application of ANNs in IVIVC has the potential to be a reliable predictive tool that overcomes some of the limitations associated with classical regression methods like lack of pattern recognition powers in analyzing multivariable data with a high degree of variation (65, 66). ANNs are also used to evaluate quantitative structure–pharmacokinetic relationship (QSPR) of beta-adrenergic antagonists demonstrating that they are helpful to predict *in vivo* results from *in vitro* experiments. In addition, the IVIVC principles have been applied in the determination of absorption parameters of salbutamol (67) in the lungs in healthy as well as asthmatic volunteers *in vivo* pharmacological studies (68, 69). ANNs are useful in predicting *in vivo* responses based on *in vitro* data studies in the pharmacological segment.

4.8.4 QUANTITATIVE STRUCTURE–ACTIVITY RELATIONSHIPS (QSAR)

QSAR principles help to correlate the structure or property descriptors of chemical or biological compounds with increasing number of neural network models (70). They are also deployed to determine the number of physicochemical properties from the existing molecular structures. All QSAR studies are based on the principle of interdependence of biological activities with respect to physicochemical parameters (71–73). Computational methods help to study the relationships between structure and function with respect to physicochemical descriptors and topological parameters.

4.8.5 QUANTITATIVE STRUCTURE–PROPERTY RELATIONSHIP (QSPR)

ANNs are useful tools to predict antimicrobial activities using topological methods (74). A case study of antimicrobial activities of quinolone (75) derivatives and determinations of their minimum inhibitory concentrations using a topological method based on chemical structural properties is a fine example. ANN principles were used in the development of four-parameter counter-propagation. QSAR model is used in predicting toxicity for freshwater alga *Pseudokirchneriella subcapitata* (76). ANNs screened antibacterial activity with example of 3-hydroxy pyridine-4-one derivatives (77) and furthermore, it has been useful in QSAR studies of antitumor activity, especially in the case studies of acridine-based derivatives (78). QSPR ANN model helps the herbicides segment to predict octanol–water partition coefficients for more than 209 chlorinated *trans*-azobenzene derivatives to determine the contaminants. QSPR ANN model was also used to calculate the polar surface area of drug molecules (79). Application of ANNs in the prediction of API solubility is enumerated in the case study comparing MLR, ANN, and SVM methods with ANN principles. Some of the AI tools used in drug delivery are mentioned in Table 4.1.

After applying AI tools in drug discovery, some of the expected outcomes are depicted in Figure 4.2.

Figure 4.3 shows the role of AI in developing pharmaceutical industry.

4.9 CONCLUSION

The drug delivery segment aids in deciding suitable excipients in formulations, monitoring, and modifying drug development process, ensuring process specification compliance, etc. DeepChem, DeepTox, DeepNeuralNetQSAR, ORGANIC, PotentialNet, Hit Dexter, DeltaVina, Neural graph fingerprint, and AlphaFold, are some of the AI in pharmaceutical drug delivery systems. QSPR tools help to resolve problems encountered in the formulation design area, such as stability issues, dissolution, and porosity (80–87). Expert systems (ES) and ANN help create a hybrid system for the development of direct-filling hard-gelatin capsules with the specifications of its dissolution profile (88).

TABLE 4.1
AI Tools and Their Application in Drug Discovery, Design, and Delivery

S. No.	Applications in Drug Discovery	Design and Delivery
1	DeepChem	Python-based MLP model of AI system to find a suitable candidate in drug discovery
2	DeepTox	Predicts toxicity of 12,000 drugs in a database and it is a kind of software in toxicology
3	DeepNeuralNetQSAR	Python-based system driven by computational tools useful in molecular activity of compounds
4	PotentialNet	Predicts binding affinity of ligands in pharmacology studies
5	DeltaVina	A scoring function for rescoring drug–ligand binding affinity
6	AlphaFold	3D structures of protein predictions
7	ORGANIC	A molecular generation tool that helps to create molecules with desired properties
8	Hit Dexter	Predicts molecules that might respond to biochemical assays using ML techniques
9	Neural graph fingerprint	Helps to determine the properties of novel molecules

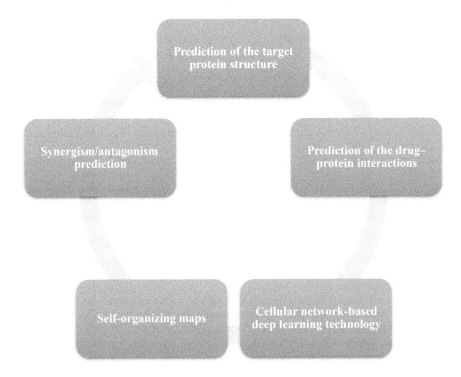

FIGURE 4.2 AI tools in drug discovery.

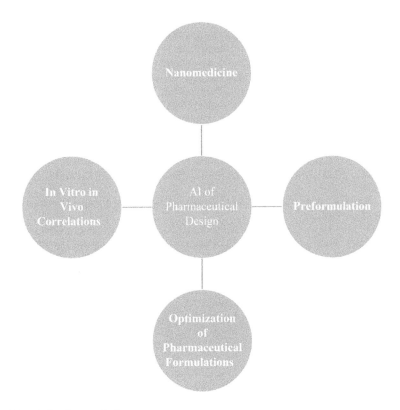

FIGURE 4.3 AI tools in pharma design.

The finite element method has been used to study the flow property of the powder on the die-filling and process during the tablet compression.

REFERENCES

1. A. N. Ramesh, C. Kambhampati, J. R. T. Monson et al. 2004. Artificial intelligence in medicine. Ann. R. Coll. Surg. Engl. 86 (5), 334–338.
2. J. Miles and A. Walker. 2006. The potential application of artificial intelligence in transport. IEE Proc. Intell. Transp. Syst. 153, 183–198.
3. Y. Yang and K. Siau. 2018. A Qualitative Research on Marketing and Sales in the Artificial Intelligence Age. MWAIS.
4. B. W. Wirtz, J. C. Weyerer, C. Geyer et al. 2019. Artificial intelligence and the public sector: Applications and challenges. Int. J. Public Adm. 42, 596–615
5. O. I. Abiodun, A. Jantan, A. E. Omolara et al. 2018. State-of-the-art in artificial neural network applications: A survey. Heliyon 4 (11), e00938.
6. M.-O. Mackenrodt (2019) Artificial intelligence and collusion. IIC Int. Rev. Intellect. Prop. Compet. Law 50, 109–113.
7. I. N. Da Silva, D. H. Spatti, R. A. Flauzino et al. 2017. Artificial Neural Networks. Cham: Springer International Publishing, p. 39.
8. L. Medsker and L. C. Jain. 1999. Recurrent Neural Networks: Design and Applications. CRC Press.
9. D. B. Kirew, J. R. Chretien, P. Bernard et al. 1998. Application of Kohonen neural networks in classification of biologically active compounds. SAR QSAR Environ. Res. 8(1–2), 93–107.
10. A. Asikainen, M. Kolehmainen, J. Ruuskanen et al. 2006. Structure-based classification of active and inactive estrogenic compounds by decision tree, LVQ and kNN methods. Chemosphere 62(4), 658–673.
11. M. Rafienia and M. Amiri. 2010. Application of artificial neural networks in controlled drug delivery systems. Appl. Artif. Intell. 24 (8), 807–820.
12. F. Montes, K. V. Gernaey, G. Sin et al. 2018. Implementation of a radial basis function control strategy for the crystallization of ibuprofen under uncertainty. Comput. Aided Chem. Eng. 44, 565–570.
13. V. K. Sutariyaa, A. Grosheva, P. Sadana et al. 2013. Artificial neural network in drug delivery and pharmaceutical research. Open Bioinform. J. 7 (Suppl.1, M5), 49–62.
14. P. Melin, J. Amezcua, F. Valdez et al. 2014. A new neural network model based on the LVQ algorithm for multi-class classification of arrhythmias. Inf. Sci. 279, 483–497.
15. B. Bajželj and V. Drgan. 2020. Hepatotoxicity modeling using counter-propagation artificial neural networks: Handling an imbalanced classification problem. *Molecules*. 25, 481.
16. B. Nagy, D. L. Galata, A. Farkas et al. 2022. Application of artificial neural networks in the process analytical technology of pharmaceutical manufacturing: A review. AAPS J. 24, 74.
17. N. Shahid, T. Rappon, W. Berta et al. 2019. Applications of artificial neural networks in health care organizational decision-making: A scoping review. PLoS One. 14(2), e0212356.
18. J. A. Anderson. 2000. *Talking Nets: An Oral History of Neural Networks*. MIT Press, ISBN 9780262511117.
19. F. Wanand and J. Zeng. 2016. Deep learning with feature embedding for compound–protein interaction prediction. bioRxiv, 086033.
20. M. AlQuraishi. 2019. End-to-end differentiable learning of protein structure. Cell Syst. 8, 292–301
21. H. S. Stoker, 2015. *Organic and Biological Chemistry*. Cengage Learning, p. 371. ISBN978-1-305-68645-8.
22. J. Jumper et al. 2021. Highly accurate protein structure prediction with AlphaFold. Nature. 596(7873), 583–589.
23. F. Wang, D. Liu, H. Wang et al. 2011. Computational screening for active compounds targeting protein sequences: Methodology and experimental validation. J. Chem. Inf. Model. 51, 2821–2828.
24. X. Xiao, J. L. Min, W. Z. Lin et al. 2015. iDrug-Target: Predicting the interactions between drug compounds and target proteins in cellular networking via benchmark dataset optimization approach. J. Biomol. Struct. Dyn. 33, 2221–2233.
25. X. Zeng, S. Zhu, W. Lu. et al. 2020. Target identification among known drugs by deep learning from heterogeneous networks. Chem. Sci. 11, 1775–1797.
26. J. Achenbach, P. Tiikkainen, L. Franke et al. 2011. Computational tools for polypharmacology and repurposing. Future Med. Chem. 3, 961–968.

27. P. Hassanzadeh, F. Atyabi, R. Dinarvand et al. 2019. The significance of artificial intelligence in drug delivery system design. Adv. Drug Delivery Rev. 151, 169–190.

28. M. Luo, Y. Feng, T. Wang, J. Guan. et al. 2018. Micro-/nanorobots at work in active drug delivery. Adv. Funct. Mater. 28, 1706100.

29. J. Fu and H. Yan. 2012. Controlled drug release by a nanorobot. Nat. Biotechnol. 30, 407–408.

30. B. Wilson and Geetha K. M. 2020. Artificial intelligence and related technologies enabled nanomedicine for advanced cancer treatment. Future Med. 15, 433–435.

31. S. Sim and N. K. Wong. 2021. Nanotechnology and its use in imaging and drug delivery (Review). Biomed Rep. 14 (5), 42.

32. D. Ho, P. Wang, T. Kee et al. 2019. Artificial intelligence in nanomedicine. Nanoscale Horiz. 4, 365–377.

33. G. M. Sacha and P. Varona. 2013. Artificial intelligence in nanotechnology. Nanotechnology 24, 452002.

34. Y. H. Zhao, M. H. Abraham, A. Ibrahim et al. 2007. Predicting penetration across the blood–brain barrier from simple descriptors and fragmentation schemes. J. Chem. Inf. Model. 47(1), 170–175.

35. C. Suenderhauf, F. Hammann, J. Huwyler et al. 2012. Computational prediction of blood–brain barrier permeability using decision tree induction. Molecules 17 (9), 10429–10445.

36. X. Mei, H.-C Lee, K. Diao et al. 2020. Artificial intelligence–enabled rapid diagnosis of patients with COVID-19. Nat. Med. 26(8), 1224–1228.

37. K. Gao, D. D. Nguyen, R. Wang et al. 2020. Machine intelligence design of 2019-nCoV drugs. bioRxiv. doi:10.1101/2020.01.30.927889.

38. S. A. Forbes, D. Beare, H. Boutselakis et al. 2016. COSMIC: Somatic cancer genetics at high resolution. Nucleic Acids Res. 45 (D1), D777–D783.

39. D. Paulz, G. Sanapz, S. Shenoyz et al. 2021. Artificial intelligence in drug discovery and development. Drug Discov. Today, 26, 80–92.

40. S. Colombo. 2020. Applications of artificial intelligence in drug delivery and pharmaceutical development. Artif. Intell. Healthc. 85–116.

41. P. Hassanzadeh, F. Atyabi, R. Dinarvand et al. 2019. The significance of artificial intelligence in drug delivery system designs. Adv. Drug Deliv. Rev. 151–152, 169–190.

42. N. Fleming 2018. How artificial intelligence is changing drug discovery. Nature. 557, S55–S57.

43. Y. Ni, W. Xiao, S. Kokot et al. 2009. Application of chemometrics methods for the simultaneous kinetic spectrophotometric determination of aminocarb and carbaryl in vegetable and water samples. J. Hazard. Mater. 168, 1239–1245.

44. M. Hasani, M. Moloudi, F. Emami et al. 2007. Spectrophotometric resolution of ternary mixtures of tryptophan, tyrosine, and histidine with the aid of principal component-artificial neural network models. Anal. Biochem. 370, 68–76.

45. L. Bai, H. Zhang, H. Wang et al 2006. Analysis of ultraviolet absorption spectrum of Chinese herbal medicine-Cortex Fraxini by double ANN. Spectrochim. Acta A 65, 863–868.

46. L. Zhu, S. H. Shabbir, E. V. Anslyn et al. 2007. Two methods for the determination of enantiomeric excess and concentration of a chiral sample with a single spectroscopic measurement. Chemistry 13, 99–104.

47. S. Agatonovic-Kustrin, I. G. Tucker, D. Schmierer et al. 1999. Solid state assay of ranitidine HCl as a bulk drug and as active ingredient in tablets using DRIFT spectroscopy with artificial neural networks. Pharm. Res. 16, 1477–1482.

48. S. Agatonovic-Kustrin, V. Wu, T. Rades et al. 1999. Powder diffractometric assay of two polymorphic forms of ranitidine hydrochloride. Int. J. Pharm. 184, 107–114.

49. S. Agatonovic-Kustrin, V. Wu, T. Rades, et al. 2000. Ranitidine hydrochloride X-ray assay using a neural network. J. Pharm. Biomed. Anal. 22, 985–992.

50. T. Baczek, A. Bucinski, A. R. Ivanov et al. 2004. Artificial neural network analysis for evaluation of peptide MS/MS spectra in proteomics. Anal. Chem. 76, 1726–1732.

51. J. E. Madden, N. Avdalovic, P. R. Haddad et al. 2001. Prediction of retention times for anions in linear gradient elution ion chromatography with hydroxide eluents using artificial neural networks. J. Chromatogr. A 910, 173–179.

52. S. C. Stefanovic, T. Bolanca, M. Lusa, S et al. 2012. Multi-criteria decision making development of ion chromatographic method for determination of inorganic anions in oilfield waters based on artificial neural networks retention model. Anal. Chim. Acta 716, 145–154.

53. T. Bolanca, S. Cerjan-Stefanovic, M. Lusa et al. 2008. Valuation of separation in gradient elution ion chromatography by combining several retention models and objective functions. J. Sep. Sci. 31, 705–713.

54. S. Y. Tham and S. Agatonovic-Kustrin. 2002. Application of the artificial neural network in quantitative structure-gradient elution retention relationship of phenylthiocarbamyl amino acids derivatives. J. Pharm. Biomed. Anal. 28, 581–590.

55. S. Agatonovic-Kustrin, M. Zecevic, L. Zivanovic et al. 1998. Application of artificial neural networks in HPLC method development. J. Pharm. Biomed. Anal. 17, 69–76.
56. G. Chen, J. Li, S. Zhang et al. 2012. A sensitive and efficient method to systematically detect two bio-phenols in medicinal herb, herbal products and rat plasma based on thorough study of derivatization and its convenient application to pharmacokinetics with semi-automated device. J. Chromatogr. A 1249, 190–200.
57. N. K. Ebube, G. Owusu-Ababio, M. Adeyeye et al. 2000. Preformulation studies and characterization of the physicochemical properties of amorphous polymers using artificial neural networks. Int. J. Pharm. 196, 27–35.
58. Y. Onuki, S. Kawai, H. Arai et al. 2012. Contribution of the physicochemical properties of active pharmaceutical ingredients to tablet properties identified by ensemble artificial neural networks and Kohonen's self-organizing maps. J. Pharm. Sci. 101, 2372–2381.
59. B. D. Glass, S. Agatonovic-Kustrin, M. H. Wisch et al. 2005. Artificial neural networks to optimize for-mulation components of a fixed dose combination of rifampicin, ionized and pyrazinamide in a micro emulsion. Curr. Drug Discov. Technol. 2, 195–201.
60. J. Takahara. 1997. Multi-objective simultaneous optimization technique based on an artificial neural network in sustained release formulations. J. Control. Release 49, 11.
61. J. Takahara, K. Takayama, K. Isowa et al. 1997. Multiobjective simultaneous optimization based on artificial neural network in a ketoprofen hydrogel formula containing oethylmenthol as a percutaneous absorption enhancer. Int. J. Pharm. 158, 203–210.
62. A. S. Hussain, X. Q. Yu, R. D. Johnson et al. 1991. Application of neural computing in pharmaceutical product development. Pharm. Res. 8, 1248–1252.
63. S. Ibric, M. Jovanovic, Z. Djuric et al. 2002. The application of generalized regression neural network in the modeling and optimization of aspirin extended release tablets with Eudragit RS PO as matrix substance. J. Controll. Release. 82, 213–222.
64. S. Ibric, M. Jovanovic, Z. Djuric et al. 2003. Artificial neural networks in the modeling and optimization of aspirin extended release tablets with Eudragit L 100 as matrix substance. AAPS. Pharm. Sci. Tech. 4, E9.
65. J. A. Dowell, A. Hussain, J. Devane et al. 1999. Artificial neural networks applied to the *in vitro–in vivo* correlation of an extended-release formulation: Initial trials and experience. J. Pharm. Sci. 88, 154–160.
66. J. V. Gobburu and E. P. Chen. 1996. Artificial neural networks as a novel approach to integrated phar-macokinetic–pharmacodynamic analysis. J. Pharm. Sci. 85, 505–510.
67. C. H. Richardson, M. de Matas, H. Hosker et al. 2007. Determination of the relative bioavailability of salbutamol to the lungs following inhalation from dry powder inhaler formulations containing drug substance manufactured by supercritical fluids and micronization. Pharm. Res. 24, 2008–2017.
68. M. de Matas, Q. Shao, C. H. Richardson et al. 2008. Evaluation of *in vitro–in vivo* correlations for dry powder inhaler delivery using artificial neural networks. Eur. J. Pharm. Sci. 33, 80–90.
69. P. Paixão, L. s. F. Gouveia, J. A. G. Morais et al. 2010. Prediction of the *in vitro* intrinsic clearance determined in suspensions of human hepatocytes by using artificial neural networks. Eur. J. Pharm. Sci. 39, 310–321.
70. K. Z. Myint, L. Wang, Q. Tong, X. Q. Xie et al. 2012. Molecular fingerprint-based artificial neural net-works QSAR for ligand biological activity predictions. Mol. Pharm. 9, 2912–2923.
71. M. Nirouei, G. Ghasemi, P. Abdolmaleki et al. 2012. Linear and non-linear quantitative structure–activity relationship models on indole substitution patterns as inhibitors of HIV-1 attachment. Indian J. Biochem. Biophys. 49, 202–210.
72. Y. Uesawa, K. Mohri, M. Kawase et al. 2011. Quantitative structure–activity relationship (QSAR) analy-sis of tumor-specificity of 1,2,3,4-tetrahydroisoquinoline derivatives. Anticancer Res. 31, 4231–4238.
73. J. C. Dearde and M. Hewitt. 2010. QSAR modeling of bioconcentration factor using hydrophobicity, hydrogen bonding and topological descriptors. SAR QSAR Environ. Res. 21, 671–680.
74. J. V. Gobburu and W. H. Shelver. 1995. Quantitative structure pharmacokinetic relationships (QSPR) of beta blockers derived using neural networks. J. Pharm. Sci. 84, 862–865.
75. J. Jaen-Oltra, M. T. Salabert-Salvador, F. J. Garcia-March et al. 2000. Artificial neural network applied to prediction of fluorquinolone antibacterial activity by topological methods. J. Med. Chem. 43, 1143–1148.
76. M. D. Ertürk, M. T. Saçan, M. Novic et al. 2012. Quantitative structure–activity relationships (QSARs) using the novel marine algal toxicity data of phenols. J. Mol. Graph. Model. 38, 90–100.
77. R. Sabet, A. Fassihi, B. Hemmateenejad et al. 2012. Computer-aided design of novel antibacte-rial 3-hydroxypyridine-4-ones: Application of QSAR methods based on the MOLMAP approach. J. Comput. Aided Mol. Des. 26, 349–361.

78. M. Koba. 2012. Application of artificial neural networks for the prediction of antitumor activity of a series of acridinone derivatives. Med. Chem. 8, 309–319.
79. A. J. Wilczynska-Piliszek, S. Piliszek, J. Falandysz et al. 2012. Use of quantitative structure–property relationship (QSPR) and artificial neural network (ANN) based approaches for estimating the octanol–water partition coefficients of the 209 chlorinated transazobenzene congeners. J. Environ. Sci. Health B 47, 111–128.
80. H. Noorizadeh, A. Farmany, M. Noorizadeh et al. 2011. Prediction of polar surface area of drug molecules: A QSPR approach. Drug. Test. Anal. 5, 222–227.
81. B. Louis, V. K. Agrawal, P. V. Khadikar et al. 2010. Prediction of intrinsic solubility of generic drugs using MLR, ANN and SVM analyses. Eur. J. Med. Chem. 45, 4018–4025.
82. V. Sutariya, A. Groshev, P. Sadana et al. 2014. Artificial neural network in drug delivery and pharmaceutical research. Open Bioinform. J. 7, 49–62.
83. P. Hassanzadeh, F. Atyabi, R. Dinarvand et al. 2019. The significance of artificial intelligence in drug delivery system design. Adv Drug Deliv Rev. 151–152, 169–190.
84. H. Zhu, 2020. Big data and artificial intelligence modeling for drug discovery. Annu. Rev. Pharmacol. Toxicol. 60, 573–589.
85. H. L. Ciallella and H. Zhu. 2019. Advancing computational toxicology in the big data era by artificial intelligence: Data-driven and mechanism-driven modeling for chemical toxicity. Chem. Res. Toxicol. 32, 536–547
86. H. C. S. Chan, H. Shan, T. Dahoun et al. 2019. Advancing drug discovery via artificial intelligence. Trends Pharmacol. Sci. 40 (8), 592–604.
87. M. Petitjean and A.-C. Camprouxet. 2015. Silico Medicinal Chemistry: Computational Methods to Support Drug Design. Royal Society of Chemistry.
88. J. C. Pereira, E. R. Caffarena, C. N. Dos Santos et al. 2016. Boosting docking-based virtual screening with deep learning. J. Chem. Inf. Model. 56, 2495–2506.

5 Role of Artificial Intelligence (AI) Drug Discovery and Development

Prabhu Subbanna Gounder and Sobana Ponnusamy
Nandha Engineering College, Erode, Tamil Nadu, India

Hemalatha Selvaraj
Nandha College of Pharmacy, Vailkaalmedu, Tamil Nadu, India

Hanan Fahad Alharbi
Princess Nourah bint Abdul Rahman University, Riyadh, Saudi Arabia

Kiruba Mohandoss
Sri Ramachandra Institute of Higher Education and
Research, Chennai, Tamil Nadu, India

Mullaicharam Bhupathyraaj
National University of Science and Technology, Muscat, Oman

5.1 INTRODUCTION

Artificial Intelligence (AI) is dubbed as the key driver of the fourth industrial revolution (1). The time and money needed to maintain the drug research pipeline are the biggest obstacles in synthetic compounds and natural bioactive-based drug discovery and development. AI can investigate and generate a lot of data, which can then be used to create and leverage knowledge. As a result, the biggest pharmaceutical corporations in the world have already started incorporating AI into their drug discovery and development studies. Figure 5.1 shows the different types of AI which includes knowledge based systems (KBS), genetic algorithms (GAs), and deep learning. It also states the types of machine learning (ML) such as support vector machine (SVM), artificial neural networks (ANN) and deep neural networks (DNN).

5.2 AI IN DRUG DISCOVERY

AI holds enormous potential for the quick drug discovery and development of a host of novel solutions, including anticancer treatments in the modern era (2). The instruments that could help with drug development include clinical research, electronic medical records, high-resolution medical imaging, and genomic analyses. Pharmaceutical and medical researchers have access to vast datasets that can be examined by cutting-edge AI systems. This chapter reviews the potential use of computational biology and AI in synthetic compounds and natural bioactive-based novel and new drug development. In the last 50 years, high-throughput screening for known disease-associated targets has dominated drug discovery. As a result, finding new drugs has become a time-consuming, costly, and generally fruitless endeavor. When used in its broadest sense, drug development refers to the

DOI: 10.1201/9781003343981-5

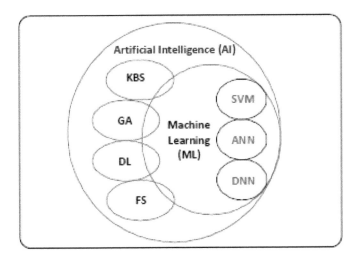

FIGURE 5.1 Different types of AI.

process of introducing a novel drug molecule into clinical practice. It covers all phases, from basic research for identifying a suitable molecular target to extensive phase III clinical studies to support the commercial launch of the drug to post-market pharmaco-surveillance and drug repurposing studies. Figure 5.2 represents that AI is being used to identify new drug targets, design new drug molecules, and predict the toxicity of the drugs.

Chemical entities that have the potential to become therapeutic agents must be identified and properly tested during the drug development process, which is a time-consuming and expensive process. Every new drug that is introduced to the market is thought to cost billions of dollars and more than ten years of labor. Thus, methods that help rapidly facilitate the drug development process are highly sought after globally (3). Figure 5.3 demonstrates that AI is being used in a variety of ways to improve the drug discovery process.

AI can be useful in a variety of ways:

a. Increasing research process agility
b. Improving predictions of therapeutic efficacy and safety
c. Increasing the chance to diversify drug pipelines

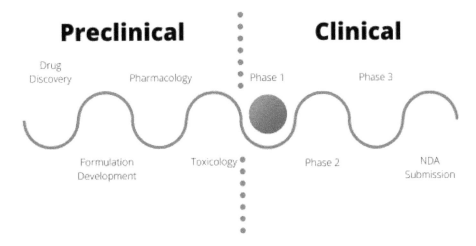

FIGURE 5.2 Preclinical and clinical studies.

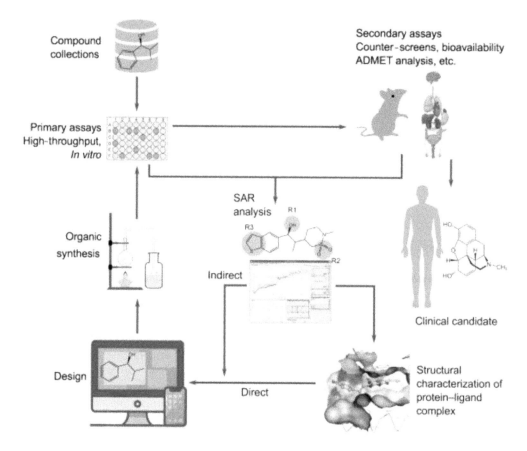

FIGURE 5.3 Advancing AI in drug discovery.

5.2.1 Critical Role Essayed by AI in Drug Discovery

Finding drugs that have a positive impact on the body, or drugs that can be used to treat or prevent a certain disease, is the main objective of drug discovery research (4). Drugs come in a wide variety of forms, but many of them are tiny molecules made through chemical synthesis. These have the capacity to bind a disease-specific target molecule. Figure 5.4 shows how AI can be used at different stages of drug discovery process. A known target is utilized to screen for small compounds that either interact with it or impact its function in cells in traditional drug discovery approaches. These methods function best for targets that are simple to drug, have a clear structure, and whose interactions inside cells are thoroughly understood. However, due to the intricate nature of cellular connections and the lack of understanding of complex cellular pathways, these methods are quite limited.

FIGURE 5.4 Drug discovery workflow.

By spotting unique interconnections and determining the relative importance of various parts of a cellular pathway, AI can get around these problems.

- To extract useful information from a large dataset, AI uses sophisticated algorithms and ML. For instance, a dataset of RNA sequencing can be used to discover genes whose expression correlates with a specific biological situation.
- AI can also be used to find substances that might bind to proteins known as "undruggable targets" because their structures are unknown. A predicted set of compounds can be quickly identified through iterative simulations of interactions between various chemicals and tiny fragments of a protein (5).

5.2.2 How Can AI Be Used in the Drug-Discovery Process?

Since the libraries utilized to find novel drug candidates are humongous, it is currently near impossible for a single researcher to review everything (6). This is where AI and ML come in to assist. AI is being used in the pharmaceutical industry to speed up the drug discovery process. The different stages of drug discovery process are shown in Figure 5.5, along with the specific applications that are being used at each stage. With the use of these advanced techniques, researchers may mine vast datasets for hidden insights and analytics.

There are various advantages of adopting this approach:

- By predicting a probable compound's qualities, only compounds with the desired properties are chosen for synthesis, which prevents time and resources from being wasted on compounds that are unlikely to be useful.
- Coming up with concepts for completely new compounds, where the "created" molecule is anticipated to have all the necessary qualities for success—which might greatly speed up the search for potent new medications.
- Reducing the need for repetitive operations, such as the examination of tens of thousands of histology images, which will save the laboratory hundreds of man-hours. Looking at the early stages of the drug discovery pipeline, these are just a few of the enormous potential benefits.

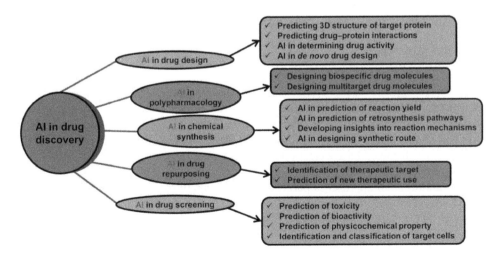

FIGURE 5.5 AI in drug discovery process. (*Source:* https://www.degruyter.com/document/doi/10.1515/ci-2022-0105/html.)

5.3   CONVENTIONAL ONCOLOGY DRUG DISCOVERY AND DEVELOPMENT

The five essential components of the traditional drug innovation work pipeline are target-differentiating proof (concepts), lead discovery, preclinical events, clinical turn of events, and administrative endorsement (7). A pharmaceutical disclosure strategy begins after assessing the impediment or implementation of a protein or pathway and depicting the potential beneficial influence. This forces the choice of a natural target, which frequently needs a lot of support before moving on to the lead drug disclosure stage. This stage entails looking for a potential improvement, such as a manageable drug like a little substance or a natural remedy. Preclinical and, if successful, clinical testing will be conducted on the drug candidate.

5.4   GENETIC ALGORITHMS (GAs) IN DRUG DISCOVERY

By using the natural selection process, the GAs are a computational technique for tackling both limited and unconstrained optimization problems (8). GAs use the genetic concepts to "generate" answers to issues. Considered are several generations of solutions, with numerous possible solutions in each generation. Data is primarily transformed from one generation to the next in three ways:

    i. In accordance with a fitness function
   ii. Through the crossover process
  iii. Through systematic mutation

   By doing this, the number of possible solutions is slightly decreased and the solutions in the following iteration improve. The fittest solutions are obtained after a few generations of this iterative process. GAs are used widely in molecular docking and QSAR. Regression analysis can be replaced by genetic function approximation. By doing several generations of QSAR analysis, they are being used for descriptor selection. Sun et al. used a mix of QSAR and GA techniques to create methyltransferase inhibitors. To achieve statistical significance, additional MLR analysis was used. Ionization potential, topological charge indices, polarizability, and the quantity of aromatic amines in a molecule are the defined descriptors (9).

5.5   EXTENSIVE APPLICATIONS OF AI IN DRUG DISCOVERY AND THEIR CLASSIFICATION

5.5.1   Target selection

An important step in the pharmaceutical advancement strategy is the approval of organic targets and the differentiating proof of those targets (10). Figure 5.6 specifies the two main genetic screens: forward genetic screens to identify genes that are involved in a particular disease or phenotype and reverse genetic screens to identify the genes that are affected by a particular drug or compound. A natural objective is a broad expression that includes, among other things, proteins, metabolites, and characteristics. It should have a clear effect and meet both industry standards and clinical and practical requirements.

   Since cheminformatics approaches permit the integration of data at several levels, by increasing the data's veracity, they offer a great deal of potential enhancements *in silico* drug design and discovery. Only a few of the algorithms that have been regularly and successfully used include bioactivity spectrum-based techniques, data mining/ML, panel docking, and searching for chemical structural similarities. The protein–ligand interaction fingerprints (PLIF) method uses a fingerprint scheme to summarize interactions between ligands and proteins; and the ligand-based interaction fingerprint (LIFt) approach uses physics-based docking and sampling methods to predict

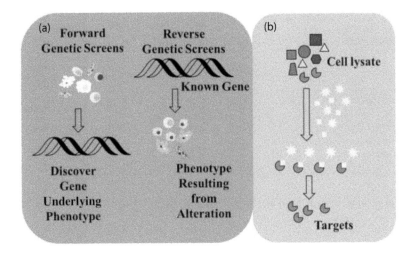

FIGURE 5.6 Reverse genetics technology.

potential targets for small-molecule drugs. Both times, compounds for the GPR17 and p38 MAP kinase were found.

5.5.2 Target Identification

Target identification focuses on determining the function of potential molecular targets (genes/proteins of a small molecule) and their significance in a disease to identify the target for a drug's effectiveness as shown in Figure 5.7 (11). This necessitates analysis of proteomics, functional and structural genomics, *in vitro* cell-based assays, and *in vivo* animal research methods. Drug Information Bank, which includes drug candidates, gene expressions, protein–protein interactions, and clinical data records, is now being analyzed by AI for the purpose of predicting therapeutic potential. For instance,

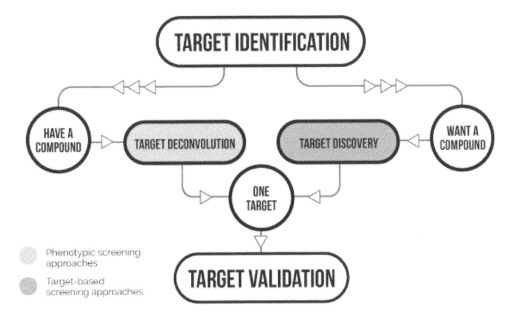

FIGURE 5.7 Target selection and validation.

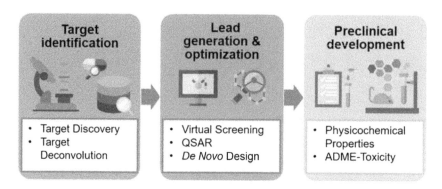

FIGURE 5.8 Steps in target identification and deconvolution.

using the "genome-wide protein interaction network, medications and their targets information," feature engineering by deep autoencoder, relief algorithm, and binary classification by Xgboost algorithm are applied to produce scores for possible targets to enable target prioritizing.

Discrete chemicals can be stored into continuous latent vector space for drug target site discovery, enabling gradient-based molecular space optimization and graph convolutional network predictions based on binding affinity and other features. There are numerous methods for identifying natural targets, which are shown in Figure 5.8. Examples of this include quality speech, proteomics, genomic research, and phenotypic screening. mRNA/protein articulation analysis is frequently employed to clarify the relationship between articulation to disease on the off chance that variations in articulation levels are connected to disturbance or movement. Targets are identified at the hereditary level by determining whether a hereditary variation and the onset of a movement are connected.

Cryo-EM microscopy data is used to train computer vision and ML algorithms in the quest to comprehend the intricate spatial 3D structure of proteins and molecular complexes (2D structure). A number of desired qualities, including pharmacokinetics, pharmacology, and toxicity profile, must be considered when choosing drug candidates. AI algorithm for medication design for a "Simplified molecular input line-entry system" string (which is a specification in the form of line notation for describing chemical species using short ASCII strings) and reinforcement learning can be successfully used. It comprises calculations of potential energy, molecular graphs with different atom or bond weights, coulomb matrices, molecular fragments or bonds, three-dimensional atomic coordinates, etc. (12).

5.5.3 Target Validation

It takes time and money to determine whether a target is crucial for a specific biological pathway, molecular function, or illness. As high-throughput screening reveals cellular responses in pertinent disease models, target validation efficiency can be considerably increased when paired with stringent data filtering and statistics (13). Figure 5.9 shows the stages of drug development which includes target identification and validation. By contrasting the network of interest with 100 random networks produced by randomly rearranging the graph while maintaining the degrees, the randomized network plug-in in Cytoscape 2.6.3 does network validation.

Genome-wide approaches and functional screens, such RNAi and CRISPR-Cas9, can be used to confirm gene function and/or gene regulatory networks. Electronic medical records and clinical trial data are now available, facilitating the recording and analysis of interindividual variability during drug administration/intervention. Free-text data from the literature can be used to find new drugs using comprehensive data mining methods in addition to molecular and clinical data.

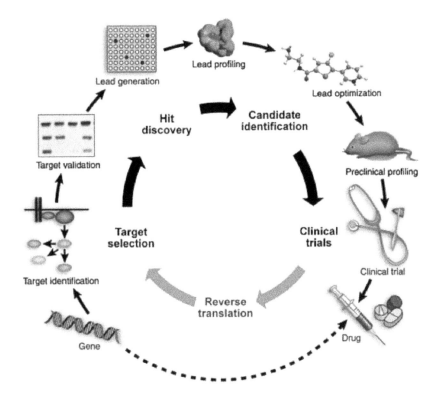

FIGURE 5.9   Stages of development of new drug.

5.6   THE CHALLENGE OF TARGET VALIDATION

Any prospective target must be examined to demonstrate that it is pertinent to the disease and that modifying it will provide the desired results, regardless of how it was discovered (14). Thus, to do so, it is necessary to have disease models and phenotypic assays that are pertinent and physiologically realistic. At present, there are access to a variety of tools and techniques for creating models; this includes genome engineering and sophisticated tissue culture techniques. These can be used to develop *in vivo* models using species like zebrafish and mice; or they can be used to create next-generation *in vitro* models using primary cells, induced pluripotent stem cells, organoids, co-cultures and microfluidic "lab on a chip" technology. For instance, all components of the lung should be able to model (including the small airway epithelium, alveoli, fibroblasts, and immune cells) and combine them in both structural and functional ways to mimic the disease phenotype in order to find more efficient ways to stop the progression of lung disease or even reverse it and repair fibrotic damage.

5.7   COMPOUND SCREENING AND LEAD OPTIMIZATION

Hits are followed by leads in the compound screening and lead optimization process (see Figure 5.10), where drug candidates are chosen using combinatorial chemistry, high-throughput screening, and virtual screening (15). The compound database for AI-based virtual screening is created by extracting large quantities from chemo-genomics libraries that are freely available to the public and contain tens of millions of compounds that have been annotated with structural information. It enables medicinal chemists to rapidly identify possible lead molecules among millions of compounds using Naive Bayes Classifiers, k-nearest neighbors, SVMs, random forests, and ANN techniques. Retrosynthesis pathway prediction is an ambitious effort to leverage AI to automate

1. Compound collections are screened against biological target (receptor, enzyme, bacteria, etc.)
2. Active compounds are identified ("Hits")
3. Hits are assessed to identify Lead (Hit to Lead)
4. Iterative rounds of design, synthesis, and evaluation are used to develop *structure–activity relationship*
5. Additional secondary assays are used to optimize compounds into clinical candidates (Lead Optimization)

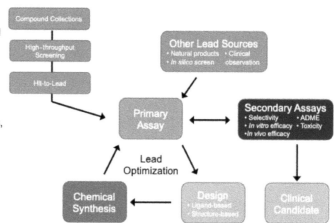

FIGURE 5.10   Flowchart of identifying lead components.

chemical synthesis with the least amount of manual labor. Here, AI and synthesis robots can be deployed. By screening out the most promising building blocks, the 3N-MCTS AI platform (which integrates three separate DNNs with Monte Carlo tree search for computer-aided organic compound synthesis) can choose only well-known reactions for the synthesis of target compounds (16).

The AI model must be taught to identify the many properties of distinct cell types swiftly and automatically in order to classify cell targets. To minimize the dimensionality of the retrieved characteristics, principal component analysis is used. AI-based techniques can be trained to categorize different cell types, such as the least-square SVM. Intelligent image-activated cell sorting devices, which evaluate optical, electrical, and mechanical cell parameters with AI-based complex DNN algorithms, are useful for accurately separating distinct cell types in the sample during the cell sorting process.

5.8   PRECLINICAL STUDIES

Preclinical studies, also known as nonclinical studies, are *in vitro* and *in vivo* laboratory testing conducted on novel medicinal substances to determine their safety and efficacy profiles (17). Preclinical process of drug discovery is shown in Figure 5.11. The time it takes to gather pertinent considerable amounts of biological data is shortened by using an unsupervised approach using clustering-based ML algorithms to analyze RNA sequencing technologies for demonstrating "molecular mechanism of action." This also reveals dozens of previously unrecognized connections between various stimuli and the cytokines they affect. *In vitro* research and preclinical pharmacokinetic investigations are used in the pharmacokinetic/pharmacodynamic ML modeling approach to forecast the "dose concentration (exposure) response relationship." Additionally, ML models are utilized in drug–dose responses to predict efficacy, which can be applied to create efficient multidrug combinations with a little amount of testing.

In carefully designed assays, Deeptox algorithm has previously been evaluated for more than 10,000 environmental chemicals and pharmaceuticals for different toxicological effects. This study of a compound's toxicological profile is a significant time- and money-consuming task. As a result, it can considerably benefit medication research by correctly forecasting a compound's toxicity. Utilizing transcriptome data from multiple biological systems and situations, deep learning algorithms are employed in "In-Silico" approaches to predict pharmacological qualities. Figure 5.12 shows the different stages of clinical trials in drug discovery process which helps to identify their properties and likelihood of the success.

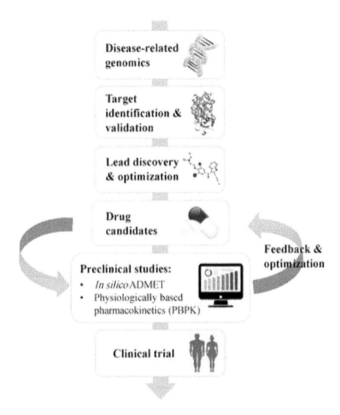

FIGURE 5.11 Preclinical studies of drug discovery.

FIGURE 5.12 Stages of drug trail.

5.9 CLINICAL TRIALS

It would be great to develop an AI tool for clinical trials that could recognize patient conditions, identify gene targets, and anticipate the effects of molecules being designed on- and off-targets. In phase II clinical trials, one such mobile application AI platform enhanced drug adherence by 25% when compared to conventional "modified directly observed therapy" (18).

Clinical trials may be conducted much more effectively in all phases, thanks to AI in risk-based monitoring, a technique that satisfies regulatory criteria while moving away from 100% source data monitoring as represented in Figure 5.13. AI can be utilized in phase II and phase III clinical trials to recognize and forecast human-relevant disease biomarkers, which can then be used to choose and enlist a particular patient population, increasing the likelihood that the trials will be successful.

5.10 FUNDAMENTAL AI

While many computational calculations can be recalled under the broad definition of AI, deep learning and AI are currently the most popular (19). Deep learning differs from traditional AI as it makes use of multiple layers, each of which makes specific estimates on the underlying data. To understand their capabilities, a few key standards must be dominated. On the other hand, contrary to what its name implies, unsupervised ML does not rely on labeled data to identify data correlations. For instance, hierarchical clustering, algorithms, and principal components analysis are used to process and sort large chemical libraries into more manageable sub-groups of related chemicals. Both classification and regression are supervised ML techniques. Classification models are employed when a complication is categorized and the enumeration result is a confined collection of worth. Regression models are used to predict a numerical value within a range of values. ML models include things like convolutional neural networks, random forests, and autoencoders.

FIGURE 5.13 Clinical trials unit.

5.11 PHASES OF A CLINICAL TRIAL

5.11.1 Phase I

Phase I trials are the initial examinations of a medication on a limited group of unharmed people. They are primarily intended to evaluate a drug's safety and tolerability, but they may also examine pharmacokinetics and, in some cases, pharmacodynamics (20). A single ascending dose (SAD) approach, in which patients are dosed in discrete cohorts, is the norm for phase I trials. A single dose of the study medicine or a placebo may be administered to every participant in a cohort. The first batch is given a very modest dose. If safety and tolerability permit, the dose is then increased in the following group. When maximum tolerance and/or maximum exposure are reached, dose increase is halted. Figure 5.14 explains how AI is being used in clinical trial process.

Multiple ascending dose (MAD) studies, which have a very similar design with cohorts and escalating dosages, are typically conducted after SAD investigations. Figure 5.15 explains the phases of clinical trial. The volunteers receive the research medicine or placebo in several doses, which is the only distinction. The multiple dosage setting frequently enables first investigations of the

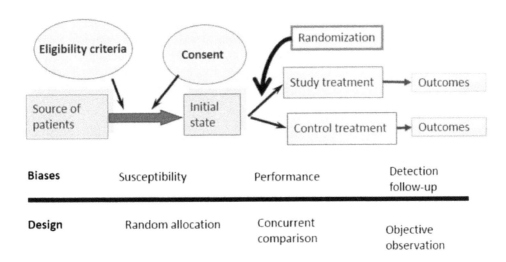

FIGURE 5.14 The clinical trial.

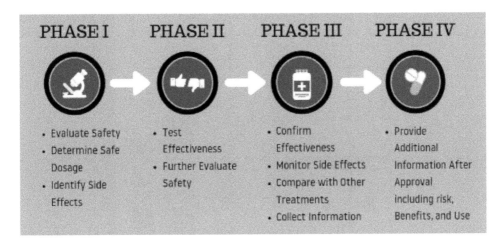

FIGURE 5.15 Phases of a clinical trial.

pharmacodynamic effects in addition to the pharmacokinetics, even though safety and tolerability are still crucial objectives. Many MAD investigations may already use patients rather than healthy individuals, depending on the risk potential and the safety and tolerability indicated by the SAD study. Having said that, it's crucial to enlist a patient population that is generally in good condition and has as few problems and comorbid disorders as possible (21).

Finally, investigations on the effects of food consumption on drug absorption are frequently carried out. Typically, two equal doses of the medicine are administered to volunteers in these experiments: one after fasting and one after eating.

5.11.2 Phase II

Larger patient populations are used in phase II trials, which are intended to evaluate the drug's effectiveness and to carry on from phase I's safety evaluations. In particular, phase II clinical trials aid in determining therapeutic doses for the extensive phase III trials. Phases IIA and IIB are frequently used to separate phase II investigations (22). Phase IIA is intended to evaluate dose needs, while phase IIB concentrates on pharmacological efficacy. The mechanism of action of the chemical has a significant impact on the precise design of phase II research. If it wasn't done during the MAD study in phase I, a proof-of-concept study should be included in phase II. A treatment study comparing various doses of the chemical with a placebo and/or an active comparator for a treatment period of 12–16 weeks is typically a crucial component of the phase II program. Figure 5.16 mentions the drug discovery from drug approval which includes preclinical and clinical research.

5.11.3 Phase III

Most of the long-term safety data are provided by phase III trials, which are randomized controlled multicenter trials (22). Phase III trials, which investigate the efficacy and safety of a new drug over 6–12 months or longer in a large patient population (a few hundred patients or more), allow assessment of the drug's overall benefit–risk relationship because they more closely resemble daily clinical life than phase I or II trials. At least in the case of metabolic illnesses, these studies are typically conducted on an outpatient basis with no in-house days and an active comparator because it would typically not be morally acceptable to subject patients to several months of placebo treatment. Phase III trials are the most expensive, time-consuming, and complex trials to design and administer, especially in therapy for chronic medical diseases, because of their scale and comparably long duration. As a result, phase III studies on medications that did not do well in phase II are frequently abandoned. About 25–30% of phase II medications move on to phase III. For the proper regulatory agencies to approve the medicine, phase IIIA trials are used (pivotal study). The submission

FIGURE 5.16 Process of drug discovery and development.

package to regulatory bodies contains the results of these studies. Phase IIIB studies are frequently carried out between submission and approval in order to gather further safety information, support publication, marketing claims (label extensions), or get the medicine ready for launch. A New Drug Application (NDA) encompassing all manufacturing, preclinical, and clinical data allows the majority of medications entering phase III clinical trials to be commercialized in accordance with FDA standards, suitable recommendations, and guidelines. Studies on drug–drug interventions (DDI) may be a component of a larger phase II or phase III trial. The DDI program, however, is dependent on the drug's DDI potential and should be negotiated with the relevant authorities. Additionally, in 2008, the US Food and Drug Administration (FDA) released a Guidance for Industry that outlined the pre-approval and post-approval requirements for the demonstration of cardiovascular safety for all new drugs created for the management of glycemic control in people with type 2 diabetes mellitus.

5.11.4 PHASE IV

Phase IV trials are also referred to as post-approval surveillance trials since they involve continuing technical support and safety monitoring (pharmacovigilance) (23). Drug approval process is mentioned in Figure 5.17. Post-marketing surveillance studies are not, however, always included in phase IV investigations. The effectiveness, cost-effectiveness, and safety of an intervention in real-world settings can be evaluated using a variety of observational designs and evaluation strategies in phase IV research.

Phase IV studies may be needed by regulatory agencies (such as a change in labeling or an action plan for risk management or minimization) or may be carried out by the sponsoring company for other factors such as competition. This can require testing of the medication on a specific new demographic (e.g., pregnant women). The goal of the safety surveillance is to find any uncommon or persistent adverse effects over a much bigger patient population and longer time frame.

5.11.5 THE POTENTIAL OF AI

AI has been used in drug disclosure since the mid-1960s. On the other hand, many big pharmaceutical companies began investing in AI in 2016, either by forming partnerships with AI start-ups

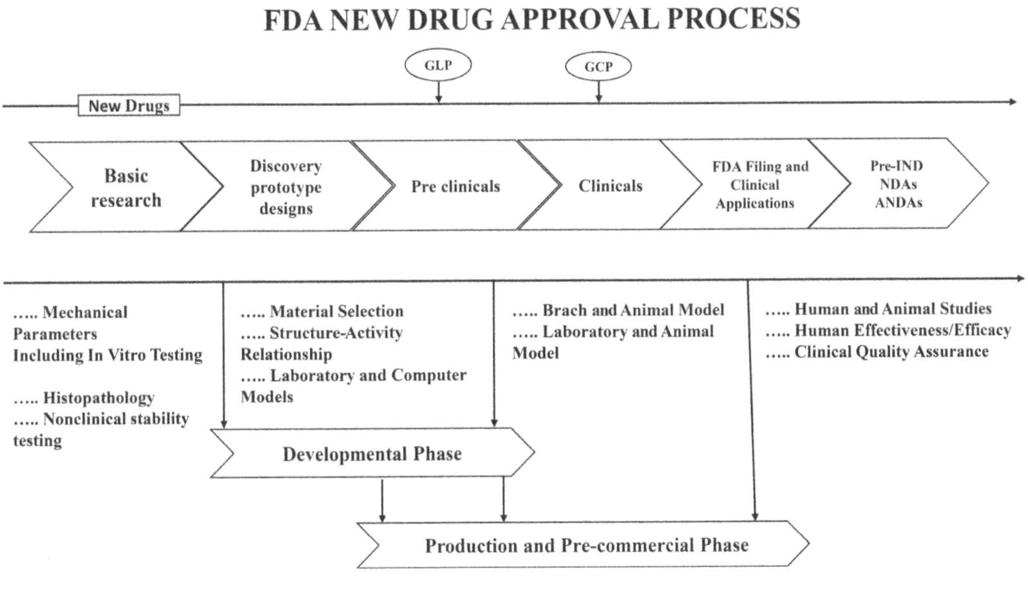

FIGURE 5.17 FDA new drug approval process.

or academic institutions or by launching their own internal AI R&D initiatives (7). This has led to an increase in new distributions covering the complete drug disclosure and advancement measure. This has encompassed anything from finding new practical targets to utilizing deep learning models to predict the features of tiny mixtures based on transcriptomics data. Man-made thinking has effectively permeated every aspect of medical discovery and advancement. Accelerating the development of the most efficient medicines and their distribution to clinics to address unmet medical needs is the main goal of AI-assisted drug discovery and development. There is a lot of potential in ML and AI. Regardless of the supplied data, AI constraints seem to have no end to newbies to the field. A multitude of uses for AI are possible and it might be able to use a model trained on photos of cats to successfully build an image of a cat, or it might be able to sit in a car and drive itself without making a mistake, or it might be able to develop a drug to treat a condition safely and effectively. AI, on the other hand, is merely a technology that can result in fresh technologies and a deeper responsiveness of the world; it cannot, therefore, solve every issue. A group of AIs that work together to improve our understanding of the drug development processes are referred to as AI in the field of drug research and development.

5.12   CASES OF AI IN DRUG DISCOVERY AND DEVELOPMENT

Each addressing a different aspect of the medication disclosure and advancement measure, a sizable amount of AI and drug revelation data are regularly given (24). Therapeutic target differentiation proof and approval, drug repurposing, finding novel mixes, and increasing R&D productivity can all be helped by AI-based medicine discovery and improvement tools. In a variety of ways, AI can reduce failures in the pipeline for traditional drug development and discovery. Objective ID and approval have advanced, thanks to simulated intelligence. This is made possible by genomic data as well as biochemical and histological details. International Business Machines (IBM) Watson identified five novel RNA-restricting proteins as prospective targets for the pathogenesis of amyotrophic lateral sclerosis, a disease for which there is currently no cure. Figure 5.18 lists some of the ways that AI can be used to improve the drug discovery and development process.

One of the biggest prospects for AI in drug discovery is medication repurposing. For instance, Donner and colleagues made another assessment of compound functioning based on high-quality

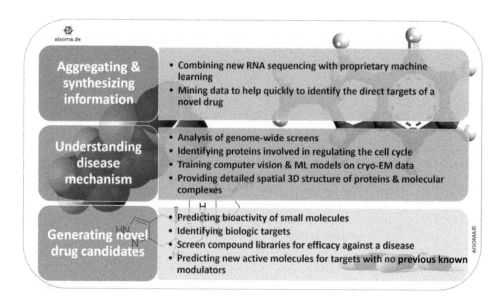

FIGURE 5.18   AI in drug discovery and development.

articulation using a transcriptomics informative index. Despite their fundamental differences, this evaluation enabled the identification of mixes that shared natural goals, revealing previously hidden utilitarian links between atoms. A competitor's tool of activity can be predicted by an AI structure; and *in vivo* security would significantly save costs. Many groups have aimed to achieve this goal and two programs, Detox and Proctor, aim to anticipate the adverse drug reactions and hypersensitivity profile of emerging synthetic drugs.

5.13 THE ADVANTAGES OF USING AI IN THE DRUG DISCOVERY MARKET

- Large libraries are used to search for potential new drugs. Even then, it is now practically impossible for lone academics to review everything; this is an area where AI and ML can be highly useful (25).
- The current pace and scope of drug discovery could change with the application of AI.
- AI does not rely on predetermined targets for drug discovery. As a result, preconceived notions and biases don't affect how drugs are developed.
- AI develops state-of-the-art drug discovery algorithms by utilizing the most recent developments in biology and computation. Due to the quick rise in processing capacity and decline in processing costs, AI can level the playing field in drug development.
- When it comes to defining relevant interactions in a drug screen, AI has a higher predictive power. As a result, by carefully setting the assay's parameters, the likelihood of false positives can be decreased.
- AI can transition drug screening from a physical lab to a virtual one, allowing for faster findings and the selection of potential targets without the need for a lot of man hours or experimental input.
- These are some possible advantages of the pipeline for drug discovery. However, as drug discovery advances and attrition rates fall, more novel treatments will eventually reach patients more quickly.

5.14 AI IN DRUG DEVELOPMENT

Figure 5.19 states the involvement of AI in drug development. The two main stages of drug development are preclinical testing through clinical trials and regulatory approval filing (26). More recently, AI has been incorporated into the development phases with the main purpose of enabling the collection, organization, and analysis of "big data" in order to enhance trial performance and regulatory approval.

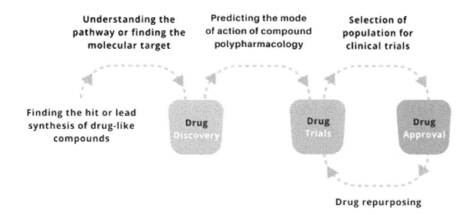

FIGURE 5.19 AI in drug development.

To conduct clinical trials, it is necessary to identify and assess clinical test locations and personnel who meet a number of requirements, including the following:

- Access to enough trial participants
- Administrative and technical support that can meet performance standards
- Capability of identifying potential trial participants from clinical records
- Clinical competency

By analyzing electronic health records to find potential participants who meet trial inclusion/exclusion criteria and integrating real-world data that may support trial performance, AI methods are being used to screen potential sites for their history of adhering to crucial enrolment criteria and compliance. ML technologies have lately become necessary due to the inclusion of digital health monitoring into clinical trial protocols, and the necessity for enhanced data processing. AI is also supplying real-time participant status updates and spotting early warning signs of potential negative outcomes. Regulation submission adheres carefully to tight standards and procedures that are evolving to include both real-world proof and data from digital monitoring systems. The usage of real-world data may also need the integration of numerous data sources with ontologies or knowledge graphs.

5.15   CHALLENGES AND OPPORTUNITIES

For achieving higher success, perhaps cheaper costs and a quicker time to market, AI techniques and technologies are naturally applied to the difficult problems of medication design and development (27). Although the outcomes to date have been more incremental than disruptive, they are nonetheless quite promising. However, some important issues that the technology by itself does not immediately address offer opportunities that can improve clinical and economic success.

5.16   LIMITATIONS OF AI IN DRUG DISCOVERY

Although AI-based approaches to drug development have a substantial impact, their capabilities and functionalities are constrained in many applications (27). The fact that many AI systems, including neural networks, are frequently criticized as being viewed as black boxes that only make an effort to establish a relationship between output and input variables based on a training dataset is one of its main drawbacks. Additionally, this instantly prompts some doubts regarding the tool's capacity to generalize circumstances that were poorly defined in the dataset. Although the solutions offered by GA approaches are quite helpful, one of their drawbacks is that they are never guaranteed to arrive to the "best" option. In the ML technique, we cannot guarantee that the ML model itself learns a few components from the data given to it, independent of what the model acquired through derivatization or heuristic reasoning. It is challenging to verify which data point from the provided data was used to train which part of an ML model. Deep learning is known to perform poorly when the data size is small to medium.

5.17   THE FUTURE OF AI IN HEALTHCARE

AI will have a significant impact on future healthcare options (27) as represented in Figure 5.20. It is the main capability underlying the development of precision medicine, which is universally acknowledged to be a critically needed improvement in healthcare. Although early attempts at making recommendations for diagnosis and therapy have been difficult, it is strongly believed that AI will eventually become proficient in that field as well. It appears likely that the majority of radiology and pathology images will eventually be reviewed by a computer given the rapid advancements in AI for imaging analysis. It will become more common to use speech and text recognition

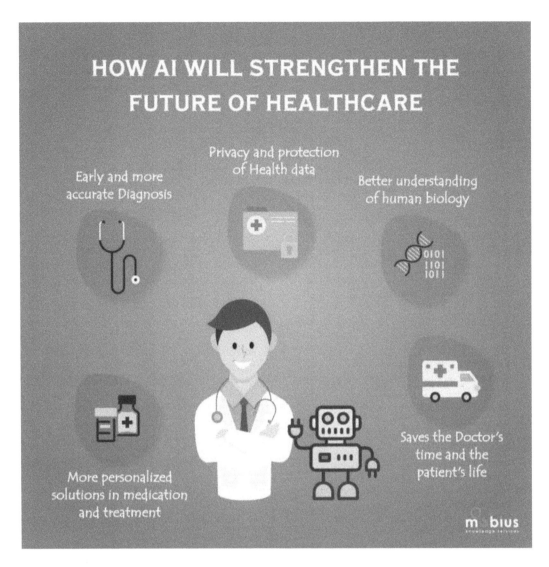

FIGURE 5.20 AI in future of healthcare.

for purposes like patient communication and clinical note transcription (28). Without determining whether the technologies will be capable enough to be beneficial, yet guaranteeing their acceptance in routine clinical practice, is the biggest hurdle for AI in various healthcare sectors.

For AI systems to be widely adopted, they need to be endorsed by regulatory bodies, integrated with EHR systems, sufficiently standardized (so that similar products function similarly), taught to clinicians, paid for by public or private payer organizations, and improved over time in the field. These difficulties will eventually be resolved, but it will take considerably longer than it will for the technology to advance (29). It is therefore anticipated that there will be modest usage of AI in clinical practice by five years and more widespread use within ten years.

Furthermore, it is increasingly obvious that AI systems will not substantially replace clinicians in patient care but rather support them. Physicians may eventually gravitate toward duties and work arrangements that make use of particularly human abilities like empathy, persuasion, and big-picture integration (30). Those healthcare professionals who refuse to collaborate with AI may end up being the only ones to lose their professions in the future.

5.18 CONCLUSION

Over the past ten years, AI in drug development has experienced a meteoric rise in popularity. Today, AI technologies are used most frequently for data management and patient selection for trials. Use is rising and is anticipated to keep rising. Future studies will look at certain use cases as to how they affect drug development effectiveness and efficiency, as well as what benefits the most.

REFERENCES

1. Paul, D.; Sanap, G.; Shenoy, S.; Kalyane, D.; Kalia, K.; Tekade, R. K. Artificial intelligence in drug discovery and development. Drug Discov Today. 2021, 26, 80–93
2. Schneider, P.; Walters, W. P.; Plowright, A. T.; Sieroka, N.; Listgarten, J.; Good-now, R. A.; Fisher, J.; Jansen, J. M.; Duca, J. S.; Rush, T. S., et al. Rethinking drug design in the artificial intelligence era.Nat. Rev. Drug Discov. 2020, 19, 353–364
3. Boström, J.; Brown, D. G.; Young, R. J.; Keserü, G. M. Expanding the medicinal chemistry synthetic toolbox. Nat. Rev. Drug Discov. 2018, 17, 709–727.
4. Pushpakom, S.; Iorio, F.; Eyers, P. A.; Escott, K. J.; Hopper, S.; Wells, A.; Doig, A.; Guilliams, T.; Latimer, J.; McNamee, C., et al. Drug repurposing: progress, challenges and recommendations. Nat. Rev. Drug Discov. 2019, 18, 41–58.
5. Tsigelny, I. F. Artificial intelligence in drug combination therapy. Brief. Bioinform. 2019, 20, 1434–1448.
6. Paananen, J.; Fortino, V. An omics perspective on drug target discovery platforms. Brief. Bioinform. 2020, 21, 1937–1953.
7. Hughes, J. P.; Rees, S.; Kalindjian, S. B.; Philpott, K. L. Principles of early drug discovery. Br. J. Pharmacol. 2011, 162, 1239–1249.
8. Pereira, D.; Williams, J. Origin and evolution of high throughput screening. Br. J. Pharmacol. 2007, 152, 53–61.
9. Kim, S. Getting the most out of PubChem for virtual screening. Expert Opin. Drug Discov. 2016, 11, 843–855.
10. Hu, Y.; Bajorath, J. Compound promiscuity: what can we learn from current data? Drug Discov. Today. 2013, 18, 644–650.
11. Yusof, I.; Shah, F.; Hashimoto, T.; Segall, M. D.; Greene, N. Finding the rules for successful drug optimisation. Drug Discov. Today. 2014, 19, 680–687.
12. Nicolaou, C. A.; Brown, N. Multi-objective optimization methods in drug design. Drug Discov. Today Technol. 2013, 10, e427–e435.
13. Sliwoski, G.; Kothiwale, S.; Meiler, J.; Lowe, E. W. Computational methods in drug discovery. Pharmacol. Rev. 2014, 66, 334–395.
14. Jiménez-Luna, J.; Grisoni, F.; Schneider, G. Drug discovery with explainable artificial intelligence. Nat. Mach. Intell. 2020, 2, 573–584.
15. Sydow, D.; Burggraaff, L.; Szengel, A.; van Vlijmen, H. W.; IJzerman, A. P.; van Westen, G. J.; Volkamer, A. Advances and challenges in computational target prediction. J. Chem. Inf. Model. 2019, 59, 1728–1742.
16. Bajorath, J. Duality of activity cliffs in drug discovery. Expert Opin. Drug Discov. 2019, 14, 517–520.
17. Oztürk, H.; Ozgür, A.; Schwaller, P.; Laino, T.; Ozkirimli, E. Exploring chemical space using natural language processing methodologies for drug discovery. Drug Discov. Today. 2020, 25, 689–705.
18. Jiménez-Luna, J.; Grisoni, F.; Weskamp, N.; Schneider, G. Artificial intelligence in drug discovery: recent advances and future perspectives. Expert Opin. Drug Discov. 2021, 16, 949–959
19. Andrade, R. J.; Chalasani, N.; Björnsson, E. S.; Suzuki, A.; Kullak-Ublick, G.A.; Watkins, P. B.; Devarbhavi, H.; Merz, M.; Lucena, M. I.; Kaplowitz, N., et al. Drug-induced liver injury. Nat. Rev. Dis. Primers. 2019, 5, 1–22.
20. Bender, A.; Cortes-Ciriano, I. Artificial intelligence in drug discovery: what is realistic, what are illusions? Part 1: Ways to make an impact, and why we are not there yet. Drug Discov. Today. 2020, 26, 511–524.
21. Bender, A.; Cortes-Ciriano, I. Artificial intelligence in drug discovery: what is realistic, what are illusions? Part 2: a discussion of chemical and biological data used for AI in drug discovery. Drug Discov. Today. 2021, 26, 1040–1052.
22. Walters, W. P.; Barzilay, R. Critical assessment of AI in drug discovery. Expert Opin. Drug Discov. 2021, 16, 937–947.

23. Rifaioglu, A. S.; Atas, H.; Martin, M. J.; Cetin-Atalay, R.; Atalay, V.; Doğan, T. Recent applications of deep learning and machine intelligence on in silico drug discovery: methods, tools and databases. Brief. Bioinform. 2019, 20, 1878–1912.
24. Korkmaz, S. Deep learning-based imbalanced data classification for drug discovery. J. Chem. Inf. Model. 2020, 60, 4180–4190.
25. David, L.; Thakkar, A.; Mercado, R.; Engkvist, O. Molecular representations in AI-driven drug discovery: a review and practical guide. J. Cheminform. 2020, 12, 1–22.
26. Bian, Y.; Xie, X.-Q. Generative chemistry: drug discovery with deep learning generative models. J. Mol. Model. 2021, 27, 1–18.
27. Xiong, Z.; Wang, D.; Liu, X.; Zhong, F.; Wan, X.; Li, X.; Li, Z.; Luo, X.; Chen, K.; Jiang, H., et al. Pushing the boundaries of molecular representation for drug discovery with the graph attention mechanism. J. Med. Chem. 2020, 63, 8749–8760, 59, 1205–1214.
28. Skalic, M.; Jiménez, J.; Sabbadin, D.; De Fabritiis, G. Shape-based generative modeling for de novo drug design. J. Chem. Inf. Model. 2019, 59, 1205–1214.
29. Rajan, K.; Zielesny, A.; Steinbeck, C. DECIMER: towards deep learning for chemical image recognition. J. Cheminform. 2020, 12, 65–70.
30. Stahl, N.; Falkman, G.; Karlsson, A.; Mathiason, G.; Bostrom, J. Deep reinforcement learning for multiparameter optimization in de novo drug design. J. Chem. Inf. Model, 2020, 59, 3166–3176

6 Artificial Intelligence in Pharmaceutical Healthcare

Prabhu Subbanna Gounder and Sobana Ponnusamy
Nandha Engineering College, Erode, Tamil Nadu, India

Hemalatha Selvaraj and Manisha Ganesh
Nandha College of Pharmacy, Vailkaalmedu, Tamil Nadu, India

Leena Chacko
Meso Scale Diagnostics LLC, Rockville, Maryland, USA

Hanan Fahad Alharbi
Princess Nourah bint Abdul Rahman University,
Riyadh, Saudi Arabia

Yoga Senbagapandian Rajamani
University of Maryland, College Park, Maryland, USA

6.1 INTRODUCTION

For a significant portion of the past ten years, Artificial Intelligence (AI) has made progress in the field of drug discovery. More than 150 small-molecule medications are currently being discovered by biotech companies employing an AI-first strategy, and more than 15 of these drugs are already in clinical trials, according to a recently published analysis (1). This AI-driven pipeline has been growing at a rate close to 40% every year.

Pharma businesses need to prepare for a future in which AI is frequently utilized in drug research given the revolutionary potential of AI. The applications are varied, and pharma businesses must decide where and how AI can most contribute value for them. New players are ramping up quickly and providing significant value. In actuality, this entails taking the necessary time to comprehend the whole impact (2). The AI-powered healthcare system is represented in Figure 6.1.

6.2 AI IN PHARMACEUTICAL INDUSTRY

It could be alluring to believe that AI can be implemented by a single team or a new tool. This is infrequently the case in our experience working with numerous businesses (3). Instead, a change in the discovery process is necessary to get the most out of AI. Companies must invest in data, technology, new competencies, and habits across the R&D organization if they want to benefit. For guidance and a road map for the future, they can look to the AI-first drug development businesses that are paving the way.

AI is a rapidly developing technology that has numerous uses in a variety of fields. Small, medium-sized, and international businesses are utilizing AI technology to improve their capacity to function intelligently in this digital environment. Figure 6.2 shows the AI revolution in pharmaceutical research.

DOI: 10.1201/9781003343981-6

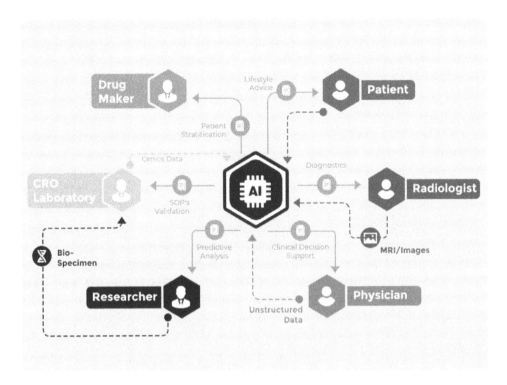

FIGURE 6.1 The AI-powered healthcare system.

AI is gaining traction in the pharmaceutical and healthcare industries, much like it has in retail, e-commerce, and manufacturing. Companies are coming up with novel solutions to some of the major problems that the pharma industry is currently dealing with by utilizing the potential of this contemporary AI (4).

AI in the pharmaceutical sector has the ability to foster innovation while also boosting output and delivering superior outcomes. Additionally, the pharmaceutical industry is benefiting from AI through developing new products.

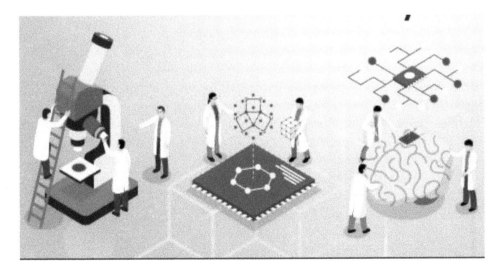

FIGURE 6.2 The AI revolution in pharmaceutical research.

Companies that offer software or services to pharma firms and are AI-native drug discovery leaders have driven much of the historical advancement. At various points throughout the value chain, these businesses employ data and analytics to enhance one or more distinct use cases. Examples include small-molecule design using generative neural networks and target finding and validation using knowledge graphs. Through collaborations or software licensing agreements, large pharmaceutical companies have been able to access these capabilities and then use them in their own pipelines (5).

Several AI-native drug discovery businesses have developed their own internal pipelines and end-to-end drug development capabilities during the past several years, ushering in a new breed of biotech company. For instance, to combine several platform technologies, Atomwise and Schrödinger established a joint venture with a shared portfolio, while Roivant Sciences bought Silicon Therapeutics. The applications of AI in pharmaceutical research is shown in Figure 6.3, which includes drug research, diagnosis, data analysis and precision medicine.

Investment has skyrocketed as a result of the shift from traditional service and software models to partnerships for asset development and pipeline development. Over the last five years, third-party investment in AI-enabled drug discovery has more than doubled annually, surpassing $2.4 billion in 2020 and exceeding $5.2 billion by the end of 2021. These figures do not include the amounts that pharmaceutical firms are spending on internal capabilities or the sums made available by IT behemoths, which have been aggressive in extending their AI investments into biology and pharmaceutical research. For instance, Nvidia has invested in the Clara suite of AI tools and applications, Alphabet just formed Isomorphic Labs based on AI advances at its DeepMind AI business, and Baidu's AI drug development arm recently inked a significant contract with Sanofi.

Although the impact of AI on conventional drug discovery is still in its infancy, it has already been demonstrated that when AI-enabled capabilities are added to a conventional process, they can significantly speed up or otherwise improve individual steps and lower the costs of conducting expensive experiments. In fact, AI algorithms have the ability to alter the majority of discovery jobs (such the design and testing of molecules) so that physical trials are only necessary to validate results (6).

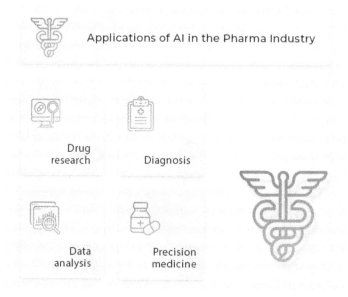

FIGURE 6.3   Applications of AI in the pharma industry.

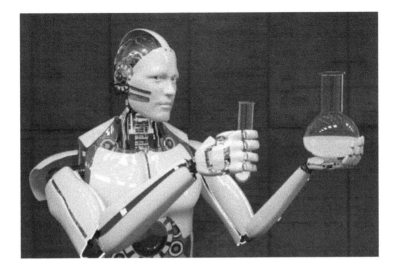

FIGURE 6.4 AI for drug safety.

6.3 THE GROWING ROLE OF AI IN PHARMACEUTICAL INDUSTRY

The application of AI in the pharmaceutical industry can aid in better patient healthcare practices, better drug development, better diagnosis and for drug safety as shown in Figure 6.4 (7). The healthcare sector includes the pharmaceutical industry as a crucial component. The industry's expansion has, however, halted during the past few years. The pharmaceutical market, in the opinion of many industry professionals, has reached saturation. The sector does, however, have optimism for technological advancement.

The pharmaceutical sector is predicted to undergo a transformation due to technologies like telemedicine, smart wearables, smart nanodevices, and AI. Of these, AI will be a key factor in advancing the sector and will find extensive use. Here is an overview of how AI will be applied in the pharmaceutical sector.

6.4 APPLICATIONS OF AI IN PHARMACEUTICAL CARE

AI is used in the pharmaceutical industry in a variety of stages of development, not just research and discovery as represented in Figure 6.5. Development is a lengthy process that begins with identifying patient's needs and ends with patient support, dosage control, and ongoing post-market research and analysis of treatment outcomes.

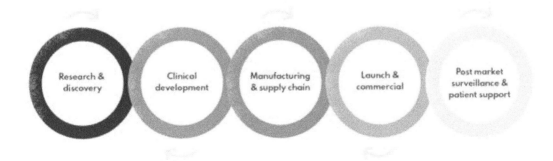

FIGURE 6.5 Product development process.

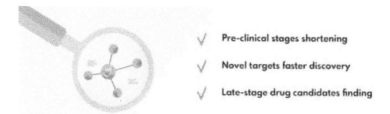

FIGURE 6.6   AI for drug discovery.

6.4.1   AI in Drug Discovery

The pharmaceutical industry's primary areas of activity are represented in Figure 6.6, such as drug research and drug development (8). The discovery step typically requires the most time because it requires thousands of studies before researchers can zero down on the ideal illness target.

The accuracy of the predictive models created for discovering unknown molecules and compounds is subject to human error. This includes improper data processing or errors in calculations. The entire process finally resulted in exorbitant expenses due to its poor development and rising failure rate.

The process of finding new drugs can be sped up by using AI. By increasing drug discovery success rates by 8–10%, applied intelligence can save the business billions of dollars. The two main applications that are trending over other pharma-based AI applications are finding compounds for drug discovery and precision medicine.

AI can be used to identify potential therapeutic candidates, create new compounds, and facilitate the synthesis of existing molecules more effectively. Figure 6.7 explains the drug discovery cycle which includes compound collections, chemical synthesis, lead compounds, design, etc. Generative modeling AI is being used by Merck KGaA and AI start-up Iktos for the quick and efficient discovery and design of novel drugs. Takeda and the drug development company Recursion have joined to examine and identify innovative preclinical candidates for rare diseases using AI (9).

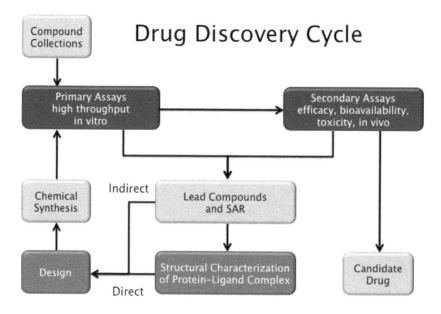

FIGURE 6.7   Drug discovery cycle.

The mechanisms of the disease are also better understood with the use of AI. Astellas Pharma, for instance, has used IMAC Lab, an AI-infused image analysis solution from LPIXEL, a pioneer in image analysis and life sciences technology. They are developing methods for selecting and managing cells for research on cell therapy and regenerative medicine. The following applications of AI show the most promising results:

Next-generation sequencing
Preclinical and early-stage drug discovery
Late-stage drug candidates
Small-molecule therapies
Novel drug design
Novel biological targets

While developing a functional drug requires a significant investment, drug development is a big industry with immense financial risks. Spending a lot of money on numerous candidate cures that ultimately fall short and get bogged down in testing or regulatory approval is not uncommon (10).

6.4.2 DRUG DEVELOPMENT

Due to human error in data processing and candidate monitoring, drug development through clinical trials carries a high probability of failure (11). Additionally, clinical trials take longer than expected, which slows down the product's commercialization. Human mistake resulting from ignorance or inattention can also lead to data interpretation errors in clinical trials, which compounds the problems that cause them to fail. Figure 6.8 represents the different stages of drug development, which includes pre-clinical and clinical research.

AI technologies and algorithms process enormous amounts of data more quickly and precisely, keep accurate records, and guarantee data transparency for clinical trials. AI shortens the entire medication development cycle through its data-driven decision-making, and because of its precision, it also increases drug approval rates and reduces loss. It can be used to optimize every step of a study, including location selection and experiment design.

FIGURE 6.8 Stages of drug development.

For instance, the clinical trial matching (CTM) feature of IBM Watson does away with the need to manually compare the enrollment criteria for clinical trials. CTM uses AI to analyze patient medical records and connect the appropriate patient with the appropriate clinical research. As a result, over the trial period (about 16 weeks), the pre-screening wait time was decreased by 78% and was automatically eliminated.

AI also makes it possible to acquire clinical data more quickly and accurately, identify and track trial applicants who are a good fit, forecast risk and toxicity, and keep track of trial candidates' drug adherence. BenevolentAI's technology and data science platform have been used by Novartis to manage clinical trial data and track candidate drug adherence. The key to efficiently developing new medications is AI (12).

6.4.3 AI IN PHARMACEUTICAL MANUFACTURING

Again, the manufacturing of drugs can take longer if it is not technologically optimized. In the pharmaceutical sector, spreadsheets and manual processes that require human interaction are still widely used in the supply chain (13). The pharmaceutical production process is explained in Figure 6.9 with the involvement of AI. In the end, it leaves potential for error because of faulty data entry and processing, conjecture, and unsuccessful results. These mistakes might result in industrial failures and significant losses.

AI can be utilized for end-to-end visibility, inventory management, demand forecasting, predictive maintenance, demand management, pharma quality control, and logistics optimization. Amgen has created an AI-powered procedure that has significantly improved its capacity to identify patterns in production irregularities and to stop them from happening again.

Through careful supply chain planning, AI improves the accuracy of the entire production process. Through predictive and prescriptive supply chain modeling, Merck KGaA has used the platform of AI company Aera Technology to support its decision-making. In the meantime, Novartis

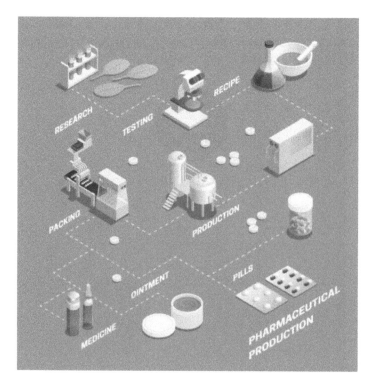

FIGURE 6.9 The pharmaceutical production process.

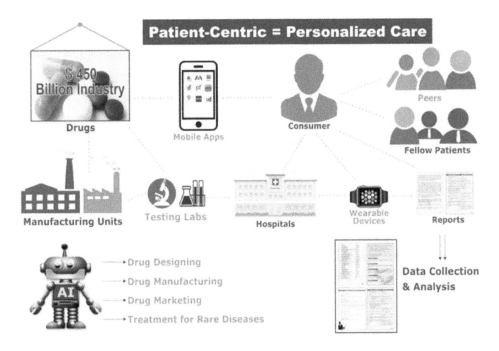

FIGURE 6.10   The personalized healthcare revolution.

has teamed with Amazon Web Services (AWS) to use cloud services to digitally revolutionize the production, supply chain, and distribution of its medications.

AI enables pharmaceutical manufacturers to improve production processes in a variety of ways:

- More consistent quality control, supporting consistently meeting critical quality attributes (CQAs)
- Shortened design phase
- Improved waste management
- Supply chain management
- Inventory management
- Improved production reuse
- Predictive maintenance

AI has the potential to increase industrial efficiency, leading to faster output and fewer waste. Reduced human involvement and data processing are primarily responsible for making this possible. The personalized healthcare revolution with the advancement of AI is shown in Figure 6.10.

6.4.4   SALES AND MARKETING

For pharmaceutical businesses, the end-to-end visibility that AI provides for medication commercialization is a significant value-addition. Pharma businesses can benefit from AI's ability to predict market access, improve marketing operations, support physician decision-making, and better coordinate product introductions (14). To enhance their sales and marketing decision-making, Pfizer Australia is creating AI-based digital analyst tools through the technical consultancy firm Complexica. Figure 6.11 represents the inputs and outputs of marketing as to how AI is being used in pharmaceutical marketing. AI algorithms can analyze large datasets of customer data, such as purchase history and browsing behavior to identify customer segments and target marketing campaigns more effectively.

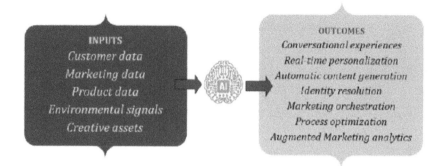

FIGURE 6.11 The inputs and outputs of marketing.

6.4.4.1 AI in Marketing

AI-powered patient engagement is a cutting-edge area of pharmaceutical marketing that not only offers advice to patients but also guarantees a pleasurable experience. For those who struggle with digestive health difficulties, Sanofi has joined with the video consulting company Babylon to provide an online AI Health Service.

In order to help patients, doctors, and nurses manage heart disease effectively, Novartis and Tencent Holdings together developed an AI nurse. To gain a deeper understanding of consumer journeys, some pharma marketers are employing AI. Andy, an AI-powered virtual assistant used by Johnson & Johnson, maps the entire process for its US clients using ACUVUE brand contact lenses, from first-time users to devoted customers. As a result, they are able to offer more individualized experiences throughout the purchasing process (15).

6.4.5 REMOTE PATIENT MONITORING AND SUPPORT

Pharma companies may also require exchangeable health data to track and interact with patients remotely after the introduction of their medications. Patient monitoring aids in the creation of more precise medications and treatments as mentioned in Figure 6.12 (16).

Applying AI to patient safety is crucial, especially in identifying potential side effects of prescription medications in real time. Pfizer has announced the imminent launch of a one-year home robot program with robotics company Catalia Health. Remote patient monitoring is shown in Figure 6.13, which monitors the patient often with the help of AI. This program will use voice interactions

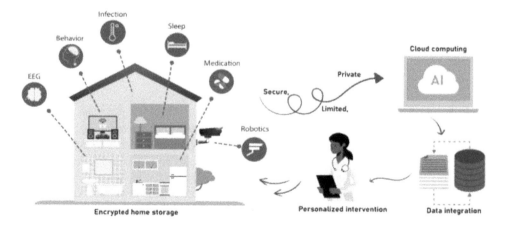

FIGURE 6.12 The power of data in patient monitoring.

FIGURE 6.13 Remote patient monitoring.

powered by conversational AI to assess a user's mood, record data, manage symptoms, and deliver useful information about prescription medications. In order to roll out a digital drug management platform across Europe, Pfizer has also partnered with SidekickHealth.

AI-enabled patient assistance programs and diagnostics can enhance patient outcomes, increase drug adherence and retention, and improve the relationship between pharma and its consumers. To create an AI tool for spotting diseases that impact low-income areas of Africa, Asia, and North and South America, MERCK KGaA worked with Johnson & Johnson. Such advancements might lower the rate of illness mortality worldwide.

6.4.6 DISEASE PREVENTION

Pharmaceutical companies can employ AI to create medications for very rare diseases like Parkinson's and Alzheimer's (2). Figure 6.14 explains the involvement of ML in remote monitoring which helps to identify and prevent rare diseases.

According to Global Genes, there aren't any additional medications available to treat and cure nearly 95% of uncommon diseases more quickly. But because of the cutting-edge capacities of AI

FIGURE 6.14 Machine learning for remote patient monitoring.

and ML, the pharmaceutical business will undergo a full transformation, which will guarantee the most cutting-edge models for early detection of dangerous diseases and better patient outcomes.

6.4.7 IDENTIFYING CLINICAL TRIALS

One of the most important pharmaceutical use cases for integrating AI into current models is identifying clinical trials. The pharmaceutical industry is increasingly using AI to extract therapeutic candidates from massive clinical datasets that are now through final clinical studies (17). Figure 6.15 explains the applications of AI in identifying clinical trials, which is considered to be the future of pharmaceutical manufacturing.

AI in the pharmaceutical industry will assist businesses in quickly evaluating tens of thousands of samples and logging information about patient responses to clinical trials. Figure 6.16 shows the life cycle of patient from clinical trials to monitoring.

Here are a few benefits of applying AI to clinical trials in the pharmaceutical industry:

- AI systems or applications assess historical clinical data.
- AI applications support the evaluation of drug responses and the monitoring of drug performance.

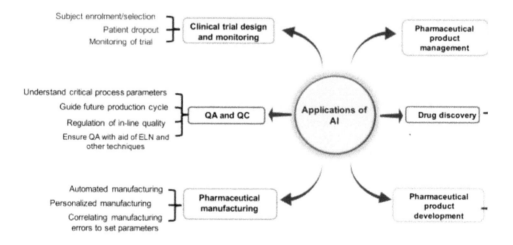

FIGURE 6.15 The future of pharmaceutical manufacturing.

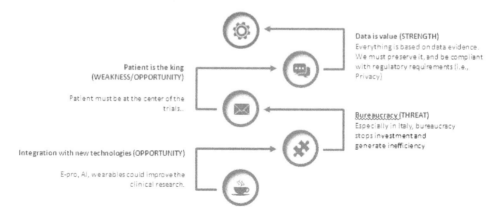

FIGURE 6.16 Patient's life cycle from prevention to recovery.

6.4.7.1 The Point of View of the Clinical Trials

AI applications for pharma will be useful for recording patients' oral text during drug trial phases, thanks to the incorporation of speech recognition technologies. It implies that AI software will capture patient's feedback.

Hence, the use of AI in clinical trials has the potential to expedite clinical trials and introduce the safest drugs into the market. It is also one of the top use cases for machine learning (ML) in pharmaceuticals. ML, deep learning, and natural language processing (NLP) technologies will be used to accurately perform speech analysis and real-time patient and drug-monitoring tasks.

6.4.8 DRUG ADHERENCES AND DOSAGE

The use of AI in pharmaceuticals and healthcare is expanding quickly in order to determine the proper dosage of medications to protect drug users. During clinical trials, AI will keep an eye on the patients and periodically recommend the appropriate dosage. Figure 6.17 represents the use of digital adherence technologies for improved healthcare outcome.

These are all important use cases for embracing AI in medicine. AI in the pharmaceutical and healthcare industries will undoubtedly speed up the automation of procedures and lead to more accuracy than previously (18).

These AI trends and application cases in pharma will help drug research and healthcare organizations guarantee efficacy throughout end-to-end production lines and offer excellent performance in front of the FDA.

Without medication adherence, no medical therapy is effective. The only way to increase a patient's chances of success is to heed physician's advice regarding their medications, food, exercise, and mental health.

Up to 60% of patients are thought to disregard medical advice, which has the impact of lowering success rates and raising expenses for treatment. Every year, nonadherence costs the United States hundreds of billions of dollars and results in thousands of fatalities; this might be easily avoided with the right technology.

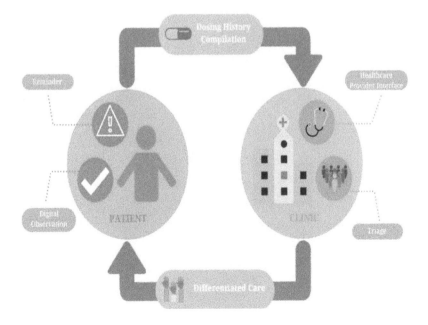

FIGURE 6.17 Digital adherence technologies for improved healthcare outcomes.

The only way to regulate adherence is through frequent in-person visits to the doctor's office due to the paucity of resources needed to provide remote care services at the patients' homes. Adherence monitoring is seeing increased adoption of AI technology. By utilizing multiple Internet-of-Things devices and centralized data collection, this can be accomplished in a variety of ways.

When a pill is ingested, ingestible sensors with RFID tags can broadcast a specific signal to a relay device and then send a signal to a cloud-based server.

While there is no way to know for sure whether a medication was really taken, a smart dispenser can keep track of how many pills are left and send reminders if necessary. On the market, there are numerous connected medication platforms, such as Pillsy, HERO, PRIA, TinyLogics, or CYCO.

Point-of-care drug assays use sophisticated "bedside" or in-clinic testing equipment to analyze urine or serum samples in order to gauge patient adherence to prescribed medications.

6.4.9 REGULATORY AFFAIRS

Another significant area of the pharmaceutical sector that can profit from adopting AI technology is regulatory affairs. The representation of regulatory affairs is given in Figure 6.18. Pharma firms must not only stay up to speed with the most recent national and international standards and laws, but also with the complicated discovery and research procedures. A regulatory team would normally collaborate with the pharmaceutical personnel to manage this extensive knowledge and ensure compliance. Even the most conscientious team, however, cannot ensure that the drug will reach the market (19).

AI can be used as a tool to centralize the information on important updates from regulatory bodies:

FDA
EMA
TGA
Health Canada
Medsafe
CHMP
PRAC

FIGURE 6.18   Regulatory affairs.

To ensure that a drug is approved, pharmaceutical companies need precise interpretation, application, and communication both inside and outside the organization. Regulatory affairs experts are also responsible for handling the approval process with regulatory agencies and preparing medications for regulatory submission. They also negotiate with regulatory agencies to guarantee that drugs are authorized.

- Finding solutions for limitations imposed by science and law
- Gathering and analyzing scientific information
- Developing plans for the company's success and the sale of medicines
- Ensuring that the drug's packaging and advertising adhere to national and international laws and standards

Advanced AI systems can analyze new and existing federal legislation pertaining to drug development procedures and give regulatory affairs personnel with timely, informative data to optimize these workflows. The AI platform may offer ongoing support after the medicine is in production, verifying compliance with marketing, legal, and technical documents.

Pharmaceutical regulatory affairs experts can receive alerts from AI about the most recent international, national, and state laws. The regulatory team may evaluate the drugs on the market and develop informed plans with the help of this dashboard.

6.4.10 PHARMACOVIGILANCE

The science and practices of medication safety monitoring are known as pharmacovigilance. These tasks include identifying, evaluating, comprehending, and avoiding negative drug effects or other potential drug-related issues.

Pharmacovigilance, as shown in Figure 6.19, entails gathering and analyzing vast volumes of data. Herbals, conventional and alternative medications, blood products, medical devices, herb vigilance, hemovigilance, and materiovigilance are now included in the program's list

FIGURE 6.19 Role of AI in pharmacovigilance.

of concerns. It's a perfect area to use deep learning algorithms and AI for advanced analytics because there are so few data points to look at and draw conclusions from. PV offers options to address classification and prediction issues, thanks to AI. This promotes efficiency and development of fresh thoughts (20).

AI-powered apps can automate the tedious and laborious activities involved in processing clinical cases, which reduces processing time and lowers total pharmacovigilance costs. Applying NLP to a large set of data, such as white papers, articles, books, or electronic medical records, would be another use of AI that might offer value by identifying unanticipated impacts of a new therapeutic product.

6.4.11 DRUG RESEARCH

A new drug's development and research are intricate processes. The process of creating a new drug takes 12 years and about US$2.6 billion. Furthermore, only 14% of the medications make it through clinical testing. Thus, it can be seen that existing approaches to medication development are rather ineffective. These inefficiencies can be greatly reduced with the aid of AI, which is shown in Figure 6.20. It may shorten the time needed for the study and creation of novel drugs. The physical and molecular properties of molecules can be predicted using AI. With AI, the forecast accuracy is likewise quite good and the procedure also shows to be time and money effective. As a result, it is possible to create novel medications fast and with improved patient outcomes (21).

The drugs' single disease-causing gene targeting is one of the main causes of medication development and research failures. It is possible to use AI to map out hundreds of disease-causing genes simultaneously. Then, drugs can be created to simultaneously target each disease-causing gene. As a result, AI can be used to develop treatments for diseases like ALS and Alzheimer's. AI is used to accelerate drug research and development that can simultaneously target all the genes.

6.4.12 DIAGNOSIS

Medical photos and scans can be interpreted using AI. The AI program then evaluates the input image against examples in its database that share a visual similarity. By precisely comparing the data included in the photos, it may find any anomalies that may be present. As a result, medical professionals may now rapidly and precisely detect patients' illnesses, thanks to AI. These steps are mentioned in Figure 6.21 that explain how AI is potential in healthcare. Additionally, they can then organize the patient's care accordingly. AI may be used to diagnose a variety of illnesses in the

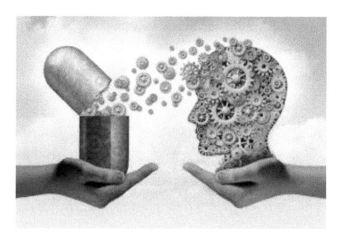

FIGURE 6.20 The future of medicine: smart pills.

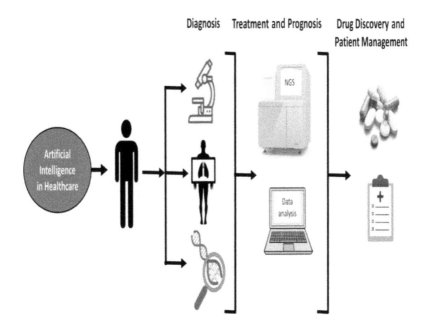

FIGURE 6.21 The potential of AI in healthcare.

healthcare sector, from hypertension to eye diseases. The most effective application of it, though, is in the treatment of fatal illnesses like cancer (22).

6.4.12.1 AI in Diagnosis

A cancer detection AI system has been created by AI start-up Lunit. A rare malignancy like airway cancer can be found and diagnosed with the aid of 3D visualization tools. With AI software, the likelihood of finding cancer has grown to 80–86%. The precision is anticipated to increase even further over time.

6.4.13 DATA ANALYSIS AND MANAGEMENT

Every day, the pharmaceutical industry generates gigabytes of data. Data may be efficiently managed, stored, and analyzed with the use of AI, as represented in Figure 6.22. Pharmaceutical businesses can benefit from AI by keeping historical records on drugs, their chemical makeup, and their use. As a result, businesses can easily access medical data as needed.

Data duplication can be avoided by using AI for data management. Duplicate data can lead to future issues, and locating the correct data might take a lot of time. Data duplication may be readily detected and removed using AI (23).

6.4.13.1 AI in Managing Medical Data

As a result, pharmacists can save a lot of time using AI when looking up historical data needed for creating new medications.

6.4.14 PRECISION MEDICINE

Large datasets can be read and analyzed by an AI model more quickly than by a human. They can be used to examine a person's and their family members' medical histories going back several generations.

FIGURE 6.22 Analyzing patient speech data.

As a result, AI can predict diseases—including hereditary ones—more accurately. Based on a person's entire medical history, AI can be used to create personalized treatments for them. Patients can be given specific or tailored medications that will hasten their recovery. Not only does AI prove useful in predicting how a patient's present medical treatment will turn out, but it can also foretell the likelihood that the patient will develop future diseases (24).

Amplion, a top provider of precision medical intelligence, recently unveiled Dx: Revenue, a piece of AI software. By comparing and analyzing data from more than 34 million data sources, this program provides pharmaceutical companies with information about the drugs they have developed. The CEO of Amplion, Chris Capdevila, asserts that "precision medicine has a problem." An unfathomable amount of information exists that has the potential to advance the development of precision medicine for patients, but it is beyond the scope of human capacity to access that information strategically, efficiently, and swiftly in order to make the best pharma partnership decisions. Our business was established in order to address this problem by offering crucial intelligence backed by evidence to help pharmaceutical and test producers make effective strategic decisions.

6.4.15 Conducting Repetitive Tasks

Data entry, medical test result analysis, and other seemingly routine and time-consuming jobs may all be completed more quickly and efficiently with AI. Doctors and other healthcare professionals will consequently have more time to concentrate on other essential and difficult tasks and communicate with patients more effectively (25).

6.4.16 Making Use of the Benefits of NLP

Since the 1950s, AI researchers have been trying to understand how human language works. NLP is a field that comprises language-related applications like speech recognition, text analysis, translation, and others. The two main methods are statistical NLP and semantic NLP. Statistical NLP is based on ML, specifically deep learning neural networks, and has helped to improve recognition accuracy recently. To learn it, you need a sizable "corpus" or body of language (26).

FIGURE 6.23 Natural language processing.

The generation, comprehension, and classification of clinical documentation and published research are the primary applications of NLP in the field of healthcare. NLP systems are able to conduct conversational AI, create reports (for example, on radiological examinations), analyze unstructured clinical notes on patients, and record patient interactions, as shown in Figure 6.23.

6.4.17 MAKING THE PROCESS OF MEDICAL CONSULTATION DIGITAL

A few AI-based apps have been created specifically to provide medical consultation based on a patient's sickness symptoms and prior medical history. The software allows users to add their symptoms. The software can then recommend a course of action after reviewing the user's medical history. The total rate of misdiagnosis is being reduced by these apps (27).

6.4.18 MAKING THE MOST OF DIGITAL NURSES' BENEFITS

Between doctor visits, digital or virtual nurses check in with the patients. AI is a fantastic technical innovation that can reduce unnecessary hospital visits. In the end, it lessens the workload for medical personnel and generates significant financial savings for the sector.

6.4.18.1 Improve Patient Care

Earlier, it was challenging to keep track of and monitor patient data while making decisions in the present. It is now simpler than ever to manage vast amounts of data and find potential treatments for various symptoms, thanks to ML. Through ML, detailed information in medical records can be used to predict actionable interventions and improve healthcare.

6.4.18.2 Predict Epidemics More Accurately

ML technology is also used to track and forecast epidemics around the world using data gathered from the Internet, social media platforms, satellites, and other well-known sources. This ML application is beneficial, particularly for nations that typically lack adequate medical infrastructure, adequate disease knowledge, and simple access to suitable treatments.

6.4.18.3   Social Media Listening

Processing the talks between patients and doctors that take place on social media is greatly aided by AI. This conversational data may be transformed into valuable insights using sophisticated ML algorithms, which can aid pharmaceutical businesses in enhancing patient experience, developing new drugs, and more.

6.5   AI IN PRIMARY AND SECONDARY DRUG SCREENING

AI has grown to be a very popular and in-demand modern technology because it saves money and time. In particular, a drug development system based on AI can both reduce and speed up time-consuming and exhausting tasks like cell classification, cell sorting, identifying small-molecule features, analyzing organic material with software, creating new material, putting assays into place, and attempting to predict the 3D shape of functional groups. A fundamental step in the drug screening process is the classification and ordering of cells utilizing AI image processing (see Figure 6.24) (28).

Numerous ML techniques that employ a variety of techniques accurately recognize photographs; but, while processing vast amounts of data, they lose their effectiveness. By contrasting the visual of the target sites with the background, the ML design is trained to recognize the target cell and its properties before being used to categorize it.

6.5.1   The Adoption of AI by Major Pharmaceutical Companies

Industry leaders have adopted AI in their workplaces after realizing its advantages (see Figure 6.25). The two major players, Pfizer and GlaxoSmithKline, have used AI in the following ways:

6.5.1.1   Pfizer

In order to speed up the development of immuno-oncology drugs, Pfizer teamed up with IBM Watson in 2016. Since then, it has partnered with numerous important institutions to accelerate the drug research and development process. The Massachusetts Institute of Technology recognized Pfizer as a member of the Machine Learning for Pharmaceutical Discovery and Synthesis Consortium in 2018. In a similar vein, it has worked with CytoReason and Concerto AI to advance drug discovery.

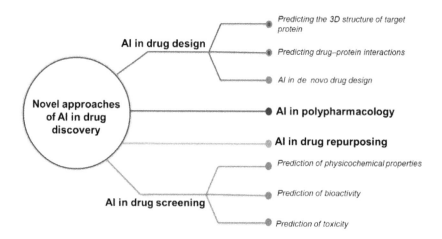

FIGURE 6.24   Approaches of AI in drug discovery and drug screening.

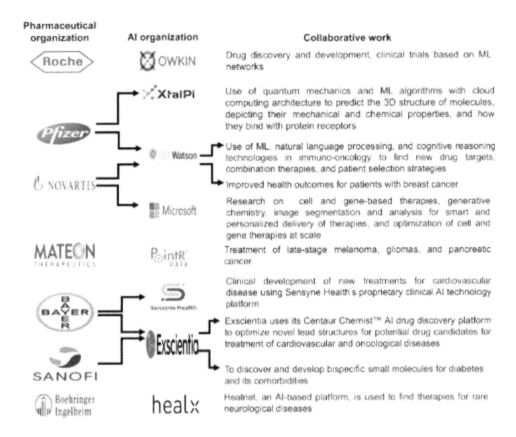

FIGURE 6.25 The adoption of AI by pharmaceutical companies.

Additionally, Pfizer revealed plans for a one-year pilot initiative in 2019 to use AI to comprehend patients' clinical experiences. Together with Catalia Health, the program was introduced. The application makes use of Mabu, a robot with AI capabilities that educates people about prescription medications (29).

6.5.1.2 GlaxoSmithKline

One of the leading pharmaceutical businesses to use AI is GlaxoSmithKline. One of the founding members of the Accelerating Therapeutics for Opportunities in Medicine (ATOM) partnership is GlaxoSmithKline. The partnership wants to change the time- and resource-intensive, high-failure drug discovery process into a rapid, patient-focused method. It has even collaborated with Google to develop implanted biomedical devices. These gadgets have the ability to alter the electric signals that travel via the body's nerves. They are able to identify abnormal or changed impulses, which are frequently present in those who have specific disorders. It has also teamed up with AI start-ups to create small-molecule treatments that can target up to ten disease-related targets.

Despite all the potential advantages AI holds, pharmaceutical businesses have just recently begun to utilize the technology. Less than 5% of healthcare businesses have adopted or invested in AI, according to a survey by HIMSS Analytics. Businesses wishing to integrate AI into their work infrastructure can join with companies that have deep knowledge in the technology to create specialized solutions for their work processes, collaborate with hospitals, or fund internal AI R&D. Most likely, AI will dominate the market in the future. AI can be used in pharmaceutical operations to save costs, streamline workflows, and—most importantly—help save lives. Businesses should get over their concerns about integrating AI and look forward to doing so in order to advance the sector.

6.6 CONCLUSION

Pharma businesses are being led toward a new era of innovation by AI. Industrywide, this technology is gradually accumulating more success stories. To treat patients with chronic obstructive pulmonary disease, GSK and Exscientia, for example, produced the first-ever AI-powered medication candidate (COPD). The medication, which has advanced to human trials, exhibits promise.

Consequently, AI technologies have enormous potential to solve the diminishing ROI in the pharmaceutical industry. The adoption of AI is being hampered by challenges like cybersecurity, transparency, and data restrictions, so AI companies are constantly attempting to improve their technologies. Despite its potential, not all pharmaceutical businesses have the funding or personnel to utilize these technologies. The demand of the future, as AI develops, is for a potential solution that can cut its cost while also simplifying its complex nature in order to increase ROI.

REFERENCES

1. Smalley, E. AI-powered drug discovery captures pharma interest. Nat. Biotechnol. 2017, 35, 604–605.
2. Hussain, A.; Malik, A.; Halim, MU.; Ali, AM. The use of robotics in surgery: A review. Int. J. Clin. Pract. 2014, 68, 1376–1382.
3. https://www.sartorius.com/en/knowledge/science-snippets/the-trending-role-of-artificial-intelligence-in-the-pharmaceutical-industry-599278, 2020.
4. Brown, N.; Ertl, P.; Lewis, R.; Luksch, T.; Reker, D.; Schneider, N. Artificial intelligence in chemistry and drug design. J. Comput. Aided Mol. Des. 2020, 34, 709–715.
5. Nadkarni, P.; Ohno-Machado, L.; Chapman, W. Natural language processing: An introduction. J. Am. Med. Inform. Assoc. 2011, 18, 544–551.
6. Spiegel, J. O.; Durrant, J. D. AutoGrow4: An open-source genetic algorithm for *de novo* drug design and lead optimization. J. Cheminfo. 2020, 12, 1–16.
7. Lamberti, M. J.; Wilkinson, M.; Donzanti, B. A.; Wohlhieter, G. E.; Parikh, S.; Wilkins, R. G.; Getz, K. A study on the application and use of artificial intelligence to support drug development. Clin. Ther., 2019, 41, 1414–1426.
8. Brazil, R. Artificial intelligence: Will it change the way drugs are discovered? Pharm J. 2017, 299, 1–10.
9. Vamathevan, J.; Clark, D.; Czodrowski, P.; Dunham, I.; Ferran, E.; Lee, G.; Li, B.; Madabhushi, A.; Shah, P.; Spitzer, M. Applications of machine learning in drug discovery and development. Nat. Rev. Drug Discov. 2019, 18, 463–477.
10. Fooladi, H. Deep Learning in Drug Discovery. *Towards Data Science*, 2020.
11. Alagappan, M., et al. A multimodal data analysis approach for targeted drug discovery involving topological data analysis (TDA). Adv. Exp. Med. Biol. 2016, 899, 253–268.
12. Chen, H.; Engkvist, O.; Wang, Y.; Olivecrona, M.; Blaschke, T. The rise of deep learning in drug discovery. Drug Discov. Today 2018, 23, 1241–1250.
13. https://stefanini.com/en/trends/news/a-guide-to-ai-in-the-pharmaceutical-industry, 2020.
14. WuXi Global Forum Team. Artificial Intelligence Already Revolutionizing Pharma. Pharmaceutical Executive; 2018.
15. Lipinski, C. F.; Maltarollo, V. G.; Oliveira, P. R.; Da Silva, A. B.; Honorio, K. M. Advances and perspectives in applying deep learning for drug design and discovery. Front. Robot. AI 2019, 108.
16. Ross, C.; Swetlitz, I. IBM pitched its Watson supercomputer as a revolution in cancer care. It's nowhere close. STAT 2017.
17. https://www.forbes.com/sites/cognitiveworld/2020/12/26/the-increasing-use-of-ai-in-the-pharmaceutical-industry/?sh=3e2bf97e4c01, 2020.
18. Bush J. How AI is taking the scut work out of health care. *Harvard Business Review*, March 5, 2018.
19. Zeng, X.; Tu, X.; Liu, Y.; Fu, X.; Su, Y. Toward better drug discovery with knowledge graph. Curr. Opin. Struct. Biol. 2022, 72, 114–126.
20. Ebejer, J.-P.; Finn, P. W.; Wong, W. K.; Deane, C. M.; Morris, G. M. Ligity: A non-superpositional, knowledge-based approach to virtual screening. J. Chem. Inf. Model. 2019, 59, 2600–2616.
21. Maltarollo, V. G.; Kronenberger, T.; Espinoza, G. Z.; Oliveira, P. R.; Honorio, K. M. Advances with support vector machines for novel drug discovery. Expert Opin. Drug Discov. 2019, 14, 23–33.
22. Zhu, Y.; Elemento, O.; Pathak, J.; Wang, F. Drug knowledge bases and their applications in biomedical informatics research. Brief. Bioinfo. 2019, 20, 1308–1321.

23. Gens, S.; Brolund, G.; Powell, S. Pursuing world class regulatory information management (RIM); Strategy, measures and priorities. Regulatory Affairs Professionals Society. 2016.

24. Bajorath J. Data analytics and deep learning in medicinal chemistry. Future Med. Chem. 2018, 13, 1541–1543.

25. Vishnoi, S.; Matre, H.; Garg, P.; Pandey, S. K. Artificial intelligence and machine learning for protein toxicity prediction using proteomics data. Chem. Biol. 2020, 96, 902–920.

26. Jiménez-Luna, J.; Grisoni, F.; Schneider, G. Drug discovery with explainable artificial intelligence. Nat. Mach. Intell. 2020, 2, 573–584.

27. Lo, Y.-C.; Rensi, S. E.; Torng, W.; Altman, R. B. Machine learning in chemoinformatics and drug discovery. Drug Discov. Today 2018, 23, 1538–1546.

28. Davenport T. H. The AI Advantage. Cambridge: MIT Press, 2018.

29. Bharatam, P. V. Computer-aided drug design. In Drug Discovery and Development. Springer, 2021, pp. 137–210.

7 Role of Machine Learning in Automated Detection and Sorting of Pharmaceutical Formulations

Ashok Kumar Janakiraman and Kushagra Khanna
UCSI University, Kuala Lumpur, Malaysia

Ramkanth Sundarapandian
Karpagam College of Pharmacy, Coimbatore, Tamil Nadu, India

7.1 INTRODUCTION

The challenge of treating complicated clinical problems using expertise and data has recently increased the digitization of data, particularly in the pharmaceutical industry. Artificial Intelligence (AI) can help by handling large amount of information with improved automation (1). It is a technology-based system which involves numerous sophisticated tools and networks that imitate human intellect but does not replace the existence of human physically. It has a high level of cognition and is involved in reasoning task such as recognizes concepts and objects, performs complex task, understands language, creates innovative ideas, and comes out with novel plans using systems and software which is able to analyze the input data and make decisions to achieve the target objectives (2). Deep learning, meta learning, neutral network, symbolic learning, generative adversarial network, reinforcement learning, and support vector machine are a few areas where AI can be efficiently employed in the drug discovery and development process (Figure 7.1).

During the golden years of AI, that is, 1956–1974, the first chatbot (a computer program that simulates human conversation, as if talking to a real person) was developed, mainly focused on algorithms to solve complex mathematic equation. Apart from that, the first machine vision learning in robots was also created in Japan in 1972. However, AI went through its first pitfall during 1974–1980 and the second pitfall during 1987–1993 due to lack of funding from the government on AI research area (3). Nevertheless, AI started to emerge again in 2010 mainly due to huge improvement on the computing power and greater data accessibility. The development of advanced and comprehensive algorithms, the arrival of low-cost graphics processors that can perform calculations in few milliseconds, and the accessibility of databases are the three main important discoveries of AI research (4).

Machine learning (ML) is one among the strategies under AI and it is an umbrella term in which it instructs the algorithm or computer program to perform intelligent behavior based on datasets. These datasets are often large, accompanied by the computational power and development in algorithm design which favor the use of MI as a powerful tool in the pharmaceutical sectors (5). Typically, it involves gathering data from multiple resources, selecting the correct ML model, collecting the results, and putting the outcomes into visual graphs (6). Currently, there are three types of ML.

Supervised learning involves the use of labeled data (models) with correct output associated with it to train the algorithms. Example is image classification.

DOI: 10.1201/9781003343981-7

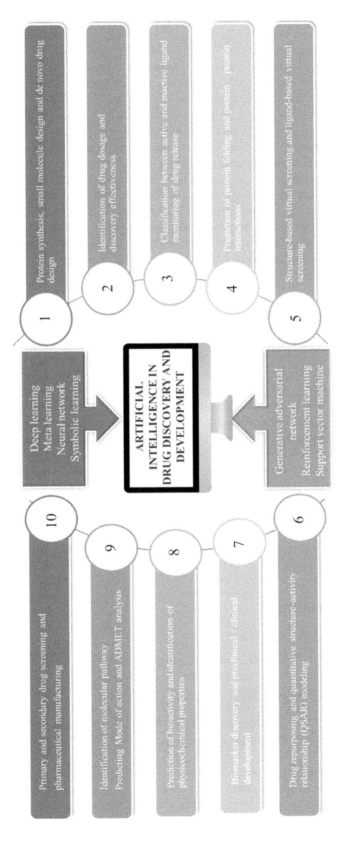

FIGURE 7.1 An illustration showing the various applications of Artificial Intelligence in drug discovery and development.

Unsupervised learning involves the algorithm that can gather the input data and visualize the models and thereby structure the data accordingly without the use of labeled data. Examples are cluster detection and pattern recognition.

Reinforcement learning involves ML's instruction to give a series of decisions in an unknown condition to achieve rewards. ML inevitably refines its approaches and improves the outcomes as it makes sufficient data. Examples of ML are Google, YouTube, etc. (7).

There are many applications of ML in pharmaceutical sciences. One of them is to identify the disease and illness and to provide individualized treatment to patients. It cannot replace the physician's roles in treating the patient, but it can enhance the medication through its large database such as early-phase clinical trials data, published results, and real-world evidence that is available and accessible to the physicians. ML can be used to determine the drug efficiency as it is based on biological factors which are related to genomics, proteomics, and metabolomics data, and these are unlikely to be obtained by using other tools (6, 7). ML methods have been employed since few years for drug discovery because of their increasing sophistication. With the recently COVID-19 pandemic, the use of ML and AI has accelerated in clinical trials to determine which medications or vaccines are effective against the coronavirus due to amplified dependence on the digital technology for patient data collection and analysis (6).

7.2   CHALLENGES AND OPPORTUNITIES

To achieve higher success, perhaps cheaper costs, and a quicker time to market, AI techniques and technologies are naturally applied to the difficult problems of medication design and development. Although the outcomes to date have been more incremental than disruptive, they are nonetheless quite promising (8). However, certain important issues that the technology by itself does not immediately address provide opportunities that can improve clinical and economic success.

Target choice is the phenotypic sufficiently specified in terms of diagnosis and stratification of the disease/condition. Few challenges reported like how thorough is patient classification, taking into account factors like clinical history, comorbidities, way of life, environment, genetics, etc. Is the aim generalizable to the patient population in the actual world, taking into account observed diversity? Can we "decode" deep learning (DL) to evaluate drug design/selection results? How are comorbidities and polypharmacy—which are present in all patients—being treated? How accurate are individual targets and responses to route modulators being modeled?

Related to clinical trials, how closely do inclusion/exclusion standards reflect patients in the real world? What effect does this have on commercialization after approval? Can illness and/or population stratification result in clinical trials that are more focused and shorter?

7.3   ALGORITHMS FOR ML

7.3.1   OVERVIEW OF ML ALGORITHMS

The assumptions made by different ML models regarding the data, the method used to create it, and the intricacy of the connections between characteristics and targets vary. In general, building an ML model necessitates striking a compromise between interpretability, robustness, and prediction abilities (9). While some models are simpler to grasp and can generate more complicated interactions, others can be too basic to account for all differences in the elemental physical relationship. Today, a huge number of ML algorithms are easily accessible through well-known libraries and software programs (10).

TensorFlow, Keras Scikit-learn, and PyTorch are some notable examples. Although a detailed analysis of supervised ML models is outside the purview of this study, we do provide a high-level summary of a few of the models that were most often used in the research here. Perhaps the most straightforward ML model is linear regression, which implies that the attributes to foresee, such

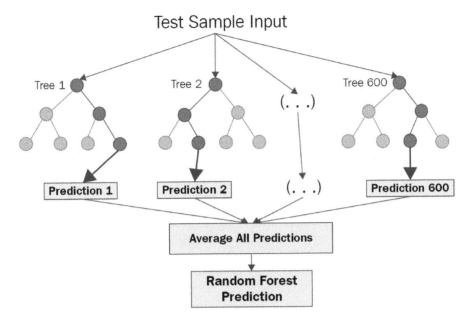

FIGURE 7.2 Random forests are an ensemble of random, uncorrelated, and fully-grown decision trees.

as solubility of active pharmaceutical ingredient (API), correlate linearly on the feed data. This model's simplicity makes it easy to understand, but it also suggests significant limits in its modeling capabilities. In general, nonlinear models perform better. While hybrid and more unusual ML model formulations have been presented, kernel-based, DL-based, and tree-based approaches are the one widely used in nonlinear ML models (11). All tree-based models are built on decision trees, which require partitioning the feature space into discrete groups. Linear approximations that link the different partitions, often known as "branches," are used to construct feature–target connections. The model's adaptability results from a training procedure that chooses the best partitions from a given dataset. Decision trees have the benefit of being interpreted, as all choices that result in a given forecast can be seen (12). Tree-based models have several well-known applications, such as random forests (RFs) and boosted trees. The latter generates predictions using many independent trees trained on subsamples of the training dataset, whereas the former trains numerous trees in a sequence to account for the variance not explained by the previous trees (Figure 7.2). In order to create predictions, kernel approaches compare fresh samples to examples from the training dataset (13). Gaussian processes (GPs), which, among the most popular kernel techniques, may provide finely calibrated measures of uncertainty on their predictions. The computational expense for training GPs increases cubically with dataset size, which is a drawback. Due to this restriction, GPs are often used with tiny datasets (e.g., less than 1,000 samples) (14). The most well-known example of sparse kernel techniques that counteract this unfavorable computational scaling is support vector machines (SVMs). There have been many additional kernel techniques created, such as relevance vector machines, radial basis function networks, and kernel ridge regression.

7.3.2 The Classification of ML Algorithms

7.3.2.1 Supervised Learning

Supervised learning is typically used to train a function that translates an input to an output using test input–output pairs. It infers a function using a variety of training samples and tagged training data. Supervised learning occurs when a task-driven strategy is used or when it is recognized that certain objectives may be achieved from a given set of inputs (15). The two most common supervised tasks are classification, which separates the data, and regression, which fits the data. In the case of a tweet or a product review, for instance, supervised learning may be used to forecast the class name or emotion of the text fragment (16).

7.3.2.1.1 Decision Tree

Each internal node in the decision tree represents a "test" of an attribute, each branch represents the result of the test, and each leaf node represents a class label. The decision tree is a type of tree-like prediction model. Its organization can aid in restoring and comprehending a problem's decision-making process (17).

7.3.2.1.2 Random Forests

Random forest, which is based on Bagging, is a key collective learning technique for regression, classification, and other similar problems. It works by supplying unlabeled samples and producing classification outcomes determined by individual trees. The decision tree's performance bottleneck is resolved by the capricious forest by adding the trapping technique to the decision tree. With the addition of randomization, it improves anti-noise performance and lowers the possibility of overfitting, demonstrating superior reproducibility and similitude in the categorization of high-dimensional data. With an unnormalized input dataset, discrete and regression issues can be effectively handled by random forests (18).

7.3.2.1.3 Naive Bayes

The Bayesian School first developed in the 1950s and 1960s. One of the statistical classification approaches, the Bayesian method, can be used to forecast the likelihood of a membership relationship and the likelihood of a specific categorization. Naive Bayes is the most basic and common type of Bayesian algorithm (19). The Naive Bayes algorithm must meet the restrictive independence notion, which states that each attribute's impact on its corresponding target variable of the specified classification is independent (Figure 7.3).

7.3.2.1.4 SVM

For a while, the SVM method was very well-liked because it nonlinearly converts the input space into a high-dimensional feature space and finds the optimal linear boundary hyperplane in the high-dimensional special space. The SVM classification algorithm's guiding concept dictates that the empirical risk be reduced when the input dataset is linearly separable. To optimize the classification interval or gap between the two categories as well as to obtain the accurately distinct data for the two categories, the best boundary plane must be identified (20). To improve the SVM's generalization skills, the classification interval must be maximized (Figure 7.4).

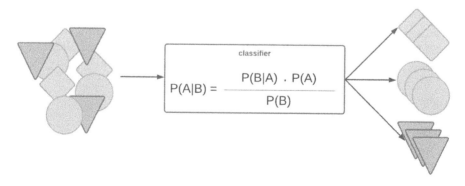

classifier

$$P(A|B) = \frac{P(B|A) \cdot P(A)}{P(B)}$$

FIGURE 7.3 Illustration of the Naive Bayes algorithm that employs the Bayes theorem.

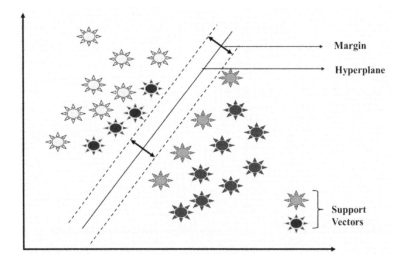

FIGURE 7.4 Illustration showing support vector machine (SVM) method that converts nonlinear boundary into a high-dimensional feature and finds the optimal linear boundary hyperplane.

7.3.2.2 Unsupervised Learning

The analysis of unlabeled datasets using data-driven unsupervised learning does not require human supervision. For experimental purposes, pertinent trend, result groupings, structure identification, and generative feature extraction, this is frequently utilized. Clustering, dimensionality reduction, density estimation, feature learning, association rule generation, anomaly detection, etc. are some of the most popular unsupervised learning tasks. For instance, clustering is the process of summarizing the data's structure and numerical values before classifying the outcomes (21). The *a priori* algorithm, K-means method, and SOM algorithm are examples of common unsupervised learning algorithms.

7.3.2.2.1 The K-means Algorithm

K-means was developed by MacQueen in the 1960s and is frequently used in cluster analysis. The principle of MacQueen K-means method is as follows:

Finding K cluster centers (a1, a2,…, ak) from the *n* data points X1, X2, …, X*n* in such a way that the squared distance between each data point and its nearest cluster center is the smallest was referred to as the objective function *W*, and its mathematical notation is as follows:

$$W_n = \sum_{i=1}^{n} \min_{1<j<k} \left| x_i - \alpha_j \right|^2$$

Since the 1970s, the K-means method has been widely used domestically and internationally in language refining (Figure 7.5), archaeology, soil, and several domains due to its simplicity in explanation, time efficiency, and suitability for processing vast amounts of data (22).

7.3.2.3 Semi-supervised Learning

With both labeled and unlabeled data, semi-supervised learning may be seen as a combination of the supervised and unsupervised methodologies previously mentioned. It falls between studying "under supervision" and "without supervision," so to speak. In instances when unlabeled data is plentiful and labeled data is in short supply in the real world, semi-supervised learning is beneficial. The primary goal of a semi-supervised learning model is to produce predictions that are better than

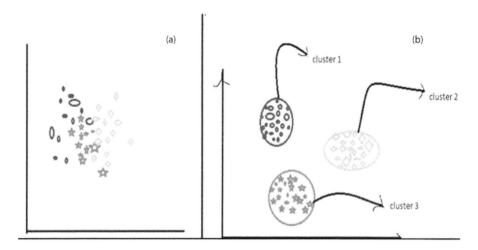

FIGURE 7.5 Classification into three different cluster categories (a) before applying K–means and (b) after applying K-means.

those obtained using only the labeled data from the model. Applications for semi-supervised learning include text classification, fraud detection, machine translation, and data labeling.

7.3.2.3.1 The Self-training Algorithm Based on the K-nearest Neighbor

The fundamental concept of the so-called K-nearest neighbor method is rather simple: It uses the training set to divide the feature space into several regions, with each sample inhabiting one of those spaces. The test sample is regarded as falling into the exact training sample category if it is located near a training sample. The K-nearest neighbor algorithm for supervised learning is discussed in detail above. K-nearest neighbor algorithms are self-training. Instead, distinct areas are divided using the feature space, which is then utilized to gradually anticipate and categorize the data category. This information is used to gradually disseminate the prediction categorization across all samples. We can get a semi-supervised learning model of K-nearest neighbor by completing self-training process.

7.3.2.3.2 The Semi-supervised Learning Algorithm Based on Divergence

The divergence-based semi-supervised learning approach starts with a technique. To categorize and identify samples taken from unlabeled test samples, training a classifier using the training set is the initial stage in cooperative training. The classifier's training set is then updated with the test samples that passed classification, and so on until all of the labels have been assigned. The advantage of this approach is that it is less affected by external factors like non-convex loss function, data quantity, and model assumptions. Additionally, this easy-to-use approach has a solid theoretical foundation and a variety of applications.

7.3.2.3.3 Semi-supervised Cluster

The clustering techniques, which are a significant category of learning techniques in semi-supervised learning, can be categorized into the following types: hierarchical clustering, partition clustering, network-based density clustering, model-based clustering, clustering high-dimensional data, constrained clustering, and outlier analysis. The method of categorizing the sample dataset into kinds with resemblance is known as "clustering." In disciplines including intrusion detection, text mining, and gene analysis, semi-supervised clustering has made significant strides. In the meantime, semi-supervised clustering has made significant advancements in GPS data edge detection, road detection, and image segmentation. As a result, semi-supervised clustering has emerged as the ML areas valued investigating in the future.

7.3.2.4 Reinforcement Learning

A method driven by the environment, reinforcement learning is a type of ML technology that enables software agents and computers to automatically analyze the best behavior in a particular context or environment to boost its efficiency. The ultimate goal of this kind of learning, which is centered on benefits and costs, is to use the information learned from environmental activists to take actions that will either increase benefits or lower costs. It is a powerful tool for creating AI models that may enhance the operational performance of complex systems like manufacturing, supply chain logistics, autonomous vehicles, robotics, etc., but it is not advised to use it to resolve straightforward or fundamental problems.

7.4 ML IN PHARMACEUTICAL FORMULATIONS

7.4.1 Conventional Oral Dosage Forms

In the 1990s, neural networks (NNs) were first utilized in pharmacological product development to predict the characteristics of instant-release (IR) oral tablets. The researcher performed experiments, variety of tablet formulations were prepared and evaluated, and the resultant data was utilized to set up NNs and/or decision trees to foresee several outcomes (e.g., dissolution rate, friability, and disintegration time) (23). This research was among the first to show the usability of ML to predict drug formulation success, although the findings were resulted in mixed results. Bourquin et al. employed NNs to anticipate the post-formulation quality parameter of tablet formulation (i.e., capping, friability, dissolution rate, disintegration time, and tensile strength at predefined time intervals) for a drug prepared using direct compression method, with relative %w/w of excipients as input factor (23).

Several researchers have examined how ML might be utilized to address a variety of issues related with the production of various types of medication formulations since the publication of Ref. (24). ML with its wide application has grown in recent years; some academics have revived the concept of ML to envisage tablet performance features based on variables. Some authors used small dataset for training, testing, and validation of the formulation. But due to small dataset and early stopping to avoid overfitting of the models, it resulted in low R2 value and high error in the validation of the samples. The authors also utilized several training procedures to reduce overfitting in their models (Figure 7.6). These included cross-validation (four times) and early quitting (25).

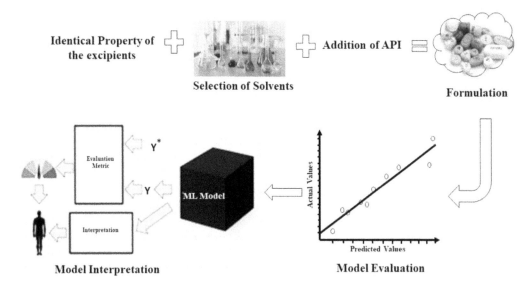

FIGURE 7.6 Illustration of AI in optimization of the formulation and best-fitted model prediction.

In general, the execution of these models was evaluated using the Pearson correlation coefficient (R), and it shows a good positive connection among anticipated and experimental findings ($R = 0.95$–0.99).

7.4.2 Advanced Oral Delivery Systems

Oral dosage forms have several clinical limitations like poor solubility or intestinal permeability. Novel oral delivery systems not only overcome these limitations but also control the rate of release of the API in the GI tract (e.g., with the help of extended-release matrix tablets) (26). The US Food and Medicine Administration (USFDA) has approved over 70% of new formulations in 2019 that were meant to be taken orally. Even with this need, there are still difficulties in creating sophisticated and effective oral formulations. The most significant of them is frequently the requirement to overcome APIs' low solubility and/or permeability, qualities that are closely connected with oral bioavailability (27). The estimation of the effects of excipients on API solubility, the prediction of API release rates from sustained-release matrix tablets, and the determination of the long-term physical stability of amorphous solid dispersions are just a few examples of the parameters for these systems that ML techniques have been developed to predict recently.

In some of the aforementioned research, ML models that have undergone cross-validation training have been utilized to accurately forecast the performance of new API–material combinations in the future, leading to the discovery of innovative medication formulations. To understand the impact of hydrotropes on the solubility of the nonsteroidal anti-inflammatory drug indomethacin, Damiati et al. utilized a NN (trained on a limited dataset by cross-validation). Delivering hydrophobic APIs, such as indomethacin, successfully typically involves overcoming poor solubility. In this work, the scientists sought to employ hydrotropes (slightly amphiphilic organic compounds) to improve indomethacin's solubility. The scientists were able to pinpoint essential characteristics that were crucial for hydrotrope-mediated solubilization of indomethacin by combining *in silico* screening and model interpretation of 16 additional hydrotropes. This research enabled the authors to determine that an "ideal" hydrotrope for IND would be a low-complexity chemical with an alkyl-substituted amide moiety, a pyridine ring acting as a hydrophobe, and a low number of hydrogen bond acceptors (28). There have also been initiatives to apply ML to enhance *in vitro* to *in vivo* correlations (IVIVC) of oral dose formulations. There are clear ethical and financial benefits to creating trustworthy ML models that can forecast *in vivo* performance, such as the drug's pharmacokinetic characteristics (29). The earliest attempts at doing this date back to the late 1990s. The scientist used nine individuals who participated in a clinical trial's pharmacokinetic profiles that were used as inputs, while *in vitro* release data for two drug formulations (i.e., USP dissolving apparatus, $n = 6$) was used as an output. The four distinct model architectures were evaluated with varying quantities of input and output attributes. In general, the goal was to predict the pharmacokinetic profile in various individuals using *in vitro* release data (30).

Additionally, ML has been incorporated toward the creation of nanomedicines. The capacity to distribute mixtures of APIs in a synergistic ratio to provide a better therapeutic effect is one of the key limitations associated with traditional medications that might be solved by formulating the nanosized advanced delivery system, such as lipid, polymeric, or lipid-based NPs. To date, the majority of research that have used ML models in the design of nanoparticles (NPs) have concentrated on employing NN to predict the characteristics of drug-loaded NPs. While the majority of these research employed tiny datasets (i.e., 50 or fewer data samples), they claim incredibly accurate prediction rates ($R2 > 0.9$) (31).

Furthermore, ML models have been developed to predict the experimental conditions necessary to prepare nanocrystals. For e.g., He et al. developed several light-gradient boosting machine models to predict the size and PDI of drug nanocrystals prepared by different techniques like high-pressure, antisolvent precipitation, and wet ball milling. In the proposed experiment to prevent model overfitting during training, the authors of this work used tenfold cross-validation with several forms of model regularization. The relevance of the 20 most pertinent elements for each optimized size prediction model was ranked important analysis feature (32).

7.5 PROGRESS AND APPLICATIONS IN PHARMACEUTICAL SCIENCE

A significant growth in the implementation of AI and ML has been seen in the pharmaceutical industry. This can be seen as the manufacturing industry is shifting toward "Industry 4.0" revolutionizing the change in the industry. Emergence of new technologies such as Internet of Things (IoT), AI, and robotics is promising in providing the enhanced output, improved manufacturing safety, better quality, better value, innovative, extra versatility, and resource efficiency. Industry 4.0 constitutes a unified, independent, and self-organizing manufacturing system. The utilization of AI Industry 4.0 however has not been widespread because it is a novel technology, expensive to implement, and regulatory hindrance. In Industry 4.0, automation processes are integrated with concepts to unify data analytics. The implementation of unified analytics has also been scarce (19).

Nowadays, the demand of productivity and superior product quality is increasing with the increased intricacies of manufacturing processes. Hence, the manufacturing practice may change and update continuously to connect human knowledge with machines. The involvement of AI in manufacturing field may bring a lot of positive impacts to the pharmaceutical industry. For example, the tools that may be involved in pharmaceutical industry such as computational fluid dynamics (CFD); Reynolds-averaged Navier–Stokes solvers technology can be used to study the effect of agitation and stress levels in different equipment such as stirred tanks, manipulating the automation of numerous pharmaceutical operations. Other than Reynolds-averaged Navier–Stokes solvers technology, the other similar systems such as large eddy simulations as well as direct numerical simulations may also comprise advanced methods to resolve the complicated flow problems in manufacturing process (33). In addition, a new computer platform called Chemical Assembly was developed, which may help with digital automation for the synthesis and manufacture of molecules by combining diverse chemical codes and utilizing it as a scripting language. Chemical Assembly has been accomplished in the process of synthesis of molecules and in manufacture of some drug compounds such as diphenhydramine hydrochloride, sildenafil, and rufinamide (34). In addition, AI technologies can be used efficiently in estimating the processing of granulation in granulators of sizes ranging from 25 to 600. Moreover, there is a method widely used in pharmaceutical industry—discrete element method (DEM). It may be used to investigate the effects of altering the blade's shape and speed, to study how powder segregates in binary mixtures, and to predict the possible path that the tablets will take throughout the coating process. Additionally, it offers analysis of how long tablet formulation stayed in the spray zone. Furthermore, there are some AI tools such as meta-classifier and tablet-classifier that can be used in governing the standard quality of the finished products. They can be used to indicate the possible errors during the process of tablet manufacturing. The patent for the AI technologies should be filed.

There is a need of balancing different kinds of parameters during the manufacturing of desired products. In order to maintain batch-to-batch uniformity and conduct quality control tests on final goods, manual intervention is required (Figure 7.7). The FDA amended the Current Good Manufacturing Practices (cGMP) by adopting a "Quality by Design" technique in order to better grasp the crucial procedure and precise standards that control the final quality of the pharmaceutical product (35). AI is applied in regulating the in-line manufacturing processes which aim to achieve the desired standard of the products. For example, the artificial neural network (ANN)-based monitoring of the freeze-drying process is implemented in this field. Back-propagation algorithms, local search, and self-adaptive evolution all are used in ANN-based monitoring of the freeze-drying process. It can be used to estimate the desiccated-cake thickness and temperature for a specific operating condition at a future time point. Besides, it can also be used to check the quality of the final products.

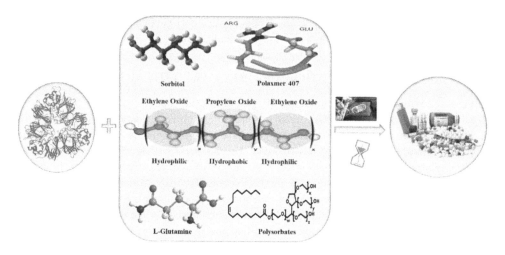

FIGURE 7.7   Illustration of AI aiding in the selection of excipients and formulation development.

Electronic Lab Notebook which is an automated data entry platform together with the intelligent techniques can be used for product quality assurance. Furthermore, total quality management expert system which consists of various knowledge discovery techniques and data mining is a valuable method in making the complicated decisions as well as creating the novel tools for intelligent quality control (36).

7.6   ML IN PHARMACEUTICAL MANUFACTURING

A shift from Industry 3.0 which was the era of automation to Industry 4.0 has resulted in tools that are better enabled by AI. This generally improves and optimizes manufacturing processes and simultaneously reduces human intervention. For example, during quality control, quality of a finished product is examined by a software to detect deviations by using images of packaging, labels, and vials. Digital twin (DT) mimics physical processes in manufacturing and uses the data to produce high-resolution models in real time to assess performance. It provides an understanding of how anomalies may have an effect on performance and yielding solutions to related risks (19).

Process controls and process performance are frequently kept apart in the pharmaceutical manufacturing environment of today. Implementing adjustments to control systems as a result has an inherent delay. The consequences of failure in the manufacturing process are a problem that has an impact on productivity and ultimately cost. This challenge can be overcome by applying AI algorithms that use ML or ANNs to detect and forecast, in real time, when measured values are drifting out of range, and weed out failure attempts early.

Sokolovic et al. demonstrated the use of programs such as INForm V.4 ANN (for NNs), FormRules V.3.32 (for neuro-fuzzy logic). and INForm V.4 GEP in improving quality of ramipril tablets. The formulation variables for ramipril tablets were refined using ANN and genetic algorithms (GA). NN programs showed that factors that affect drug stability may be acceptable as long as it is within the design space. This shows the flexibility of ramipril tablet in post-approval changes. These AI-based programs enable optimization of process parameters, hence reducing complexity of drug manufacturing. With the use of computational intelligence (CI), ANN, and Cubist, it allows forecasting of end result of various granulation processes, presents best result in optimization of twin-screw wet granulation, and foresees end result of roll compaction, respectively (37).

A study found the prospects of AI as a mechanism to aid in scale-up of high shear wet granulation process in pharmaceutical manufacturing. This study describes the flexibility of AI in predicting the operation of translating laboratory-scale information to full-scale manufacturing of the high-shear granulation process in mixers with geometric differences or operating conditions. A model was developed to identify ideal endpoint of mixer granulators with various scales ranging from 25 L to 600 L, thereby speeding scale-up while preserving geometric, dynamic, and kinematic comparability. The high-predictability model was developed using neuro-fuzzy logic and gene expression programming (GEP). Neuro-fuzzy logic technology allows identification of critical variables in the process and refines response noise, whereas GEP devises a polynomial equation that forecasts impeller power values. This innovation may provide a deeper process understanding and enhance the productivity of process transfer (38).

PAT facilitates decision-making based on a myriad of data in order to fulfill quality by design (QbD). Roggo et al. studies the DL strategies which aim to provide a better comprehension of the manufacturing process monitoring and develop a strategy for an unconventional monitoring design of a continuous manufacturing line. Deep NN analyzes and studies from noisy PAT values. It can be used to refine interpretations of PAT results when basic approaches of denoising were insufficient (39).

Gams et al. developed a novel machine-learning formula using C4.5 ML algorithm particularly for process analysis in situations when there is data scarcity, enhancing operation and quality. This method of analysis allows the operator to have an interactive simulation when changes to parameters in a manufacturing process is tested. It interprets data from previous production runs and predicts possible outcome. These plausible outcomes are presented to the operator to aid in making a more informed decision. It is a particularly suitable when there are small datasets available (40).

Polypharmacy which is the use of multiple medication at one time has been increasing dramatically. It is associated with potential adverse drug reaction and compliance threatening patient safety. 3D printing is a technology that combines different APIs into a unit dosage form at a determined patient-specific amount. It is a technic to "sandwich" multiple APIs at precise locations. The conventional way is a manual trial-and-error approach; however, this method would not be feasible with larger numbers of components which will produce uncontrollably large number of permutations. A computational algorithm would be beneficial to deal with large dataset in selecting the 3D tablet designs that meet patient-specific dissolution behavior. This personalized medicine innovation is designed in such a way that drug release matches desired dissolution profiles of the patient. Grof et al. demonstrated the use of evolutionary algorithm (EA) which simplifies the search for a particular internal structure of a multicomponent tablet that meets the specific dissolution curve. When combined with patient characteristics (required dissolution profile) data, this algorithm could be used in the development of the personalized medicine (41).

7.7 APPLICATIONS OF ML IN FORMULATION DEVELOPMENT

7.7.1 PREDICTION OF THE TARGET PROTEIN STRUCTURE

Protein is a basic compound for the development of a particular disease. However, occasionally they could be overexpressed. It is essential to forecast the structure of the target protein in order to create an effective treatment regimen and achieve selective targeting of a disease. Each protein has a unique 3D shape encoded from the amino acid sequence with one dimension (1D) which determines the biological mechanism of protein, including how it works and what it does. Again, before synthesizing any drug molecule, it requires the determination of correct target in order to provide optimal and individualized therapy. In the field of structure-based drug discovery, AI plays a major role in predicting the 3D protein structure followed by predicting the effect after a particular compound interacts with the target protein. For example, DeepMind's AlphaFold, an AI-based platform, assists in predicting the 3D target protein structure. They participate in the estimation and examination of the separation between adjacent pairs of amino acids and the associated angles of the peptide bond (17). A new accomplishment in 2020, Alpha Fold has identified five SARS-CoV-2 targets that had received insufficient research by estimating the structure of these targets (42). Targets include the Nsp2, Nsp4, Nsp6, and papain-like proteinase as well as the SARS-CoV-2 membrane protein (5).

7.7.1.1 Predicting Bioactivity/Drug–Protein Interactions

The efficacy and effectiveness of a drug molecule relies on its ability to bind to the receptor protein of the target. In order for a drug molecule to be able to produce the desired therapeutic response or outcome, it must first interact with the target protein and demonstrate affinity for it. From here, ML plays a role in measuring and predicting the binding affinity of a potential drug by considering any features or the association between the drug and its target (43). For instance, AI can be used to predict and analyze the binding affinity by discovering the geometric binding site properties as well as the noncovalent bonding patterns. AI platform such as SMILES and extended connectivity fingerprint can predict the drug features and thereafter can be used to calculate the tendency of a certain molecule to interact with their target site (17).

7.7.2 NP-MEDIATED DRUG DELIVERY

Nanotechnology has been used in pharmaceutical industry for decades because NPs have the ability to cross biological barriers, ensuring the effective delivery of drugs and other compounds such as biologics, including small interfering RNA (siRNA) and proteins, and is more specific to certain biomarkers in individual patients (44). The use of NPs improves the stability of biologics and solves the problem of low cellular penetration of genetic payloads.

However, NP features such as surface shape and charge, size, chemistry, and drug release profile are important (Figure 7.8).

If minute changes occur throughout the process and composition of formulation, the qualities of the nanomedicine can be altered, resulting in undesirable effects. Therefore, manufacturing process of NPs is important and better control is needed (45). The use of AI in the development of a system to assure the quality control of ideal NP synthesis improves the efficiency of siRNA and medication delivery, in addition to its use as a customized medicine tool. LAMMPS and GROMACS 4 are two examples of software that may be used to investigate the impact of internalization using surface chemistry on the developed NPs (34).

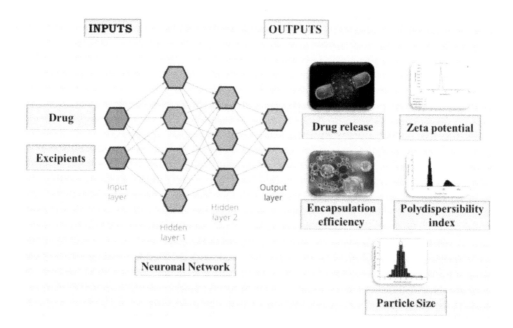

FIGURE 7.8 Illustration of the study to predict physicochemical properties and performance of the drug-loaded NPs (*in vitro* drug release, particle size, encapsulation efficiency, zeta potential, and polydispersity index), using the input features.

The application of AI in the nanomedicines includes ML. The relationship between a NP's shape and its characteristics, as well as how it interacts with the cells and tissues it is intended to target, can be better understood with the help of ML. Using ML, Puzyn et al.'s experimental evaluation forecasts the cytotoxicity of 17 distinct metal oxide NPs to *Escherichia coli* induced by the negative effects of the NPs' characteristics (44).

AI also plays an important role when there is insufficient data available and the algorithm approaches can be used. Baghaei et al. conducted a study using ML algorithm in order to reduce the prediction error from 28.0% to 2.93% for the particle size; and from 19.4% to 2.99% for initial burst release of polylactic-*co*-glycolic acid (PLGA) NPs (45). In this context, ANN which is an ML algorithm is used since it does not need huge sets of data for the study (Figure 7.9). This has demonstrated that integrating ANN into NP formulation and its evaluation offers a better substitute for the studies than the conventional computational measurements (44).

Pathologists and AI specialists must work together to improve AI systems with accurate results employing automated robotic systems for particle development in order to hasten the development of nanomedicine. Drug development for nanomedicine will be quicker and less expensive if AI-driven data processing and analysis are used, and it aids in screening of molecules and its application at both lab and industry levels (46).

Nanorobotics are also known as nanoides, nanobots, and nanomites. Generally, nanorobots are invisible to naked eye with the size of 20–200 nm constructed by manipulating macro- or micro-devices or by self-assembly on preprogrammed templets or scaffolds. Possible uses of nanorobots in medicinal fields include in detection and treatment of cancer, diagnosis and treatment of diabetes, surgery, cryostasis, gene therapy, and others (47). Nanorobots can carry and deliver the drugs to specific tissue and offer individualized treatments which increase the efficacy and decrease the aspect consequences since it has special sensors to detect the target cells. However, there are some disadvantages of nanorobotics such as issue can be raised up if anyone inhale the small debris of nanorobots since its debris are very small, it is expensive, and it is complex in manufacturing (48).

Nanorobotics can be used in the targeted drug delivery across blood–brain barrier (BBB). Nanorobots is a programmed NP to theranostic delivery across BBB (49). With the ML in AI,

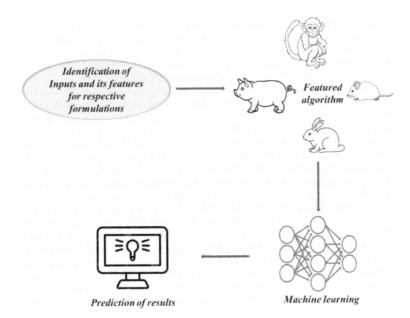

FIGURE 7.9 Illustration of the study showing macromolecule release from PLGA NPs and ML model used to predict release of the drug from the formulation.

the chemotherapeutic crossing BBB can be predicted. Due to the complex structure of the brain, the degree of side effects of drugs, and the inadequate precision of BBB model, the experimental on the theranostic agent crossing the BBB always takes a very long process which is up to decade with a very low success rate (49). By using the ML algorithms, high-dimensional data can be converted into a lower dimensional vector of coordinates for each molecule includes nearest neighbors, SVM, deep NN, and random forest. Application of appropriate algorithms can predict the BBB permeability with high accuracy and the prediction accuracy is improved continuously with deep NN model which is a four-layer multilayer perceptron (50).

7.7.3 Parenteral Formulations

Besides oral formulations, parenteral drug delivery systems are commonly applied in healthcare settings. Parenteral formulations are preparations intended for direct administration into the blood vessels, organs, and tissues. These formulations are generally given intravenously (IV), intramuscularly (IM), or subcutaneously (SC). The preparation of parenteral formulations is a high-risk activity due to its strict requirement for sterility. ML could be used to improve the manufacturing of the parenteral formulations.

ML has been attempted in the prediction of the stability profile of freeze-dried hydrocortisone powder for injection. During the stability study, the prediction of the degradation products from the dynamic neural network (DNN) was shown to be superior over the multiple regression analysis (MRA) model in terms of application usage and result handling (51). A multilayer perceptron (MLP), a class of deep ANN, was also used to aid in the prediction of the stability profile of esomeprazole powder for injection, a heat-, oxygen-, and acid-sensitive molecule. The effects of the pH value and storage duration were used to predict the performance of the esomeprazole assay and four main impurities. The MLP successfully predicted the storage condition that would stabilize the pH value of the reconstituted esomeprazole product and ensure a safe and quality product over the intended storage period (52). ML was further extended in the prediction of patient adherence to self-injectable medications. ML models trained on the data extracted from 7,697 smart sharp bins from HealthBeacon could predict and improve medication adherence in targeted patients, and long short-term memory (LSTM) has been shown to be a good predictive model (53).

7.7.4 Prediction of Drug-release Behavior in Vitro

Drug release behavior is the core and focus of drug development. It serves as a preclinical predictive model to the *in vivo* release behavior. The drug release behavior is influenced by the critical material attributes and critical process parameters such as compression pressure as well as tablet shape and size. The current method of obtaining the required release behavior involves extensive laboratory studies using specific apparatus such as the USP dissolution testers and a series of physiochemical methods. Moreover, the *in vitro* release study of a sustained-release formulation over an extended period is inconvenient. Despite the availability of accelerated release testing *in vitro*, the study is limited by the diverse responses due to the different methods used. Hence, this calls for the ML model to assist in the prediction of drug release behavior in a cost- and time-effective manner.

The ANN model has been used to predict the dissolution profile in various studies. Notable study includes determination of the release profile of extended-release anhydrous caffeine tablets via analysis of the near-infrared spectroscopy (NIR) and Raman spectroscopy. In the study, ANN provided a better prediction of the dissolution profile with a lower root mean square error (RMSE) as compared to the conventional dissolution method. An ANN that could predict the influence of critical process variables on the release behavior of 3D-printed diazepam tablets was also developed. The ANN predicted and observed results demonstrated precise prediction with a similarity factor f2. Additionally, ANN has been used to optimize the drug dissolution profile of extended-release diclofenac sodium pellets as well as predict the difference in dissolution profiles of two formulations: sustained-release tablets and fast-disintegrating films. Other ML models, such as the Elman neural networks (ENNs), demonstrate applicability in predicting drug dissolution profiles. ENN, a typical recurrent NN, has been used to describe the drug release curves of tablets and demonstrated applicability in different tablet types irrespective of hydrophilic or hydrophobic drugs. From the studies, ANN could determine the dissolution behavior based on critical material attributes, critical formulation attributes, and critical process parameters (54).

7.7.5 Pseudo-prospective Study

The study of pseudo-predictive performance of trained ML models mimics real-life applications of the model, thereby providing a more realistic estimation of its performance. For this type of study, large datasets are used in training, while validation and testing are done using a pseudo-prospective approach. The ML models will only be evaluated and validated with out-of-sample data that is recorded at a much later stage (55).

This study has been used to evaluate the model employed to predict early circulatory failure in the intensive care unit as well as the diagnosis and prognosis of meningiomas. Additionally, this approach estimated the performance of ML models in predicting sepsis in the general ward 6 hours ahead of clinical onset (56). From the study, the models were able to identify patients requiring cultures and anti-infectives; however, they were not reflective in terms of actual sepsis onset. This highlighted the potential challenges in translating retrospective data generated by ML models into actual prospective usage (57).

The main goal of IVIVC is to predict drug plasma concentration–time profiles by connecting the pharmacokinetic process *in vivo* to the drug dissolution behavior *in vitro* (often the dissolution rate or degree). When describing the performance of the product, it can lessen the strain for each phase and offer direction and support for medication development (such as reducing animal testing), production adjustments (such as replacing bioequivalence), and management and supervision (such as the quality standard establishment of dissolution rate). IVIVC models frequently take a linear route. We are aware that ANNs applied to an IVIVC have more benefits than some traditional regression approaches when it comes to data processing (30, 31).

7.8 ML AND THE PHARMACEUTICAL INDUSTRY

ML offers advantages in terms of reduction in cost and human error, as well as improvements in speed of drug to market and drug performance. Over the last ten years, there have been increasing number of pharmaceutical companies that either collaborated or acquired AI technologies for drug discovery, clinical research, diagnosis, data analysis, and predictions and discovery of novel medications (Table 7.1) (58).

7.9 REGULATORY AND RECOMMENDATION INSIGHTS

The FDA released a discussion paper on April 2, 2019, titled "Proposed Regulatory Framework for Modifications to Artificial Intelligence/Machine Learning (AI/ML)-based Software as a Medical Device (SaMD): Discussion Paper and Request for Feedback," in which it outlines its rationale for a potential method of premarket review for changes to software driven by AI and ML (59, 60).

In this hypothetical approach, the FDA would anticipate manufacturers' commitments to openness and real-world performance monitoring for AI and ML-based software used in medical devices, as well as periodic updates to the FDA on the modifications made in accordance with the protocol for algorithm change and preapproved prespecifications. Numerous modifications or alterations that could affect these software's users (including healthcare workers, patients, etc.) have been examined (61). Changes or alterations such as those involving the inputs, training with fresh datasets, or AI/ML architecture are few examples. An entire product life cycle regulatory approach

TABLE 7.1

Partnerships between Pharmaceutical and AI Companies and Their Areas of Collaboration

Pharmaceutical Company	AI Partner	Area of Collaboration
	 Generate: Biomedicines	To discover and create protein therapeutics for five clinical targets for several treatment areas and modalities
 AstraZeneca		To discover and develop new therapies for chronic kidney disease (CKD), idiopathic pulmonary fibrosis (IPF), systemic lupus erythematosus (SLE). and heart failure (HF)
Janssen Johnson & Johnson	Benevolent^AI	To license the rights to develop, manufacture. and commercialize clinical-stage novel drug candidates
NOVARTIS		To discover new potential uses of the current cancer drugs by investigating indications and responder groups for oncology drugs currently in clinical development
BAYER	turbine ada	To develop new applications for existing cancer drugs by predicting how cancer types respond to the drugs
To improve health outcomes by provision of personalized health care.
 Biogen	GENOMICS	To discover novel targets for multiple sclerosis via genetic assessment
 Pfizer	IBM Watson Health	To identify new drug targets, combination therapies, and patient selection strategies in immuno-oncology
To develop remote monitoring device for patients with Parkinson's disease
Bristol Myers Squibb		To design and optimize cardiovascular drug clinical trials
SANOFI	 OWKIN	To discover and develop new treatments for different types of cancer such as lung cancer, breast cancer, and multiple myeloma
Roche	PathAI	To develop an embedded image analysis program based on PathAI's algorithms, which will run through Roche's cloud-based uPath enterprise software
Boehringer Ingelheim	healx	To identify rare neurological disorder candidates
 gsk GlaxoSmithKline	 Exscientia	To discover new small molecules
 Takeda	 Numerate	To identify candidates for oncology, central nervous system (CNS), and gastroenterology

(TPLC) was also made available by the FDA for AI/ML-based SaMD in order to evaluate and monitor software products during their entire life span, from premarket development through commercial performance (39, 62).

7.10 CONCLUSIONS AND PERSPECTIVES

ML technologies may be used to generate and characterize both standard and nonconventional medication formulations. We have given some instances of how these methods might be applied to solve problems that arise naturally while developing new medication formulations. Pharmaceutical drug development pipelines have clearly been transformed by ML technologies in recent years, and this integration into the healthcare system is gradually gaining greater ground. Pharmaceutical scientists may be able to significantly accelerate the development of novel drug formulations by using ML techniques to generate low-cost predictions of the effects of API-material combinations on drug formulation parameters. According to the assessment, a more thorough integration of ML models into pharmaceutical sciences will give us the instruments required for the development of drug products based on trial and error, to focus our experimental work on the most promising material candidates, and to switch to a more effective data-driven formulation development process. It's important to understand the prospect for ML integration in designing of formulation. Generally, the ML in pharmaceutical sciences should not only be seen as a tool to speed up efforts but also as a doorway to the discovery of novel drugs, inventive formulations, and new information. These factors suggest that ML is ideally positioned to change the way pharmaceuticals are developed.

REFERENCES

1. Subanya B, Rajalaxmi RR (2014) Feature selection using artificial bee colony for cardiovascular disease classification. 2014 International Conference on Electronics and Communication Systems. https://doi.org/10.1109/ECS.2014.6892729
2. Vyas M, Thakur S, Riyaz B, Bansal KK, Tomar B, Mishra V (2018) Artificial intelligence: The beginning of a new era in pharmacy profession. Asian J Pharm 12:72–76
3. Collins C, Dennehy D, Conboy K, Mikalef P (2021) Artificial intelligence in information systems research: A systematic literature review and research agenda. Int J Inf Manage 60:102383. https://doi.org/10.1016/j.ijinfomgt.2021.102383
4. Das S, Dey R, Nayak AK (2021) Artificial intelligence in pharmacy. Indian J Pharm Educ Res 55:304–318
5. Nichols JA, Herbert Chan HW, Baker MAB (2019) Machine learning: Applications of artificial intelligence to imaging and diagnosis. Biophys Rev 11:111–118
6. Prasad P (2020) Influence of machine learning on pharma industries. Pharmacol Pharm Reports 1–5
7. Kolluri S, Lin J, Liu R, Zhang Y, Zhang W (2022) Machine learning and artificial intelligence in pharmaceutical research and development: A review. AAPS J. https://doi.org/10.1208/s12248-021-00644-3
8. Wuest T, Weimer D, Irgens C, Thoben KD (2016) Machine learning in manufacturing: Advantages, challenges, and applications. Prod Manuf Res 4:23–45
9. Bharadwaj PKB, Kanagachidambaresan GR (2021) Pattern recognition and machine learning. In: Prakash KB, Kanagachidambaresan GR (eds). Programming with TensorFlow. EAI/Springer Innovations in Communications and Computing. Springer, pp 105–144
10. Seeger M (2004) Gaussian processes for machine learning. Int J Neural Syst 14:69–106
11. Cortes C, Vapnik V (1995) Support-vector networks. Mach Learn 20:273–297
12. Zhang Z, Cui P, Zhu W (2022) Deep learning on graphs: A survey. IEEE Trans Knowl Data Eng 34:249–270
13. Duvenaud D, Maclaurin D, Aguilera-Iparraguirre J, Gómez-Bombarelli R, Hirzel T, Aspuru-Guzik A, Adams RP (2015) Convolutional networks on graphs for learning molecular fingerprints. In: Adv. Neural Inf. Process. Syst 2224–2232
14. Yang K, Swanson K, Jin W, et al (2019) Analyzing learned molecular representations for property prediction. J Chem Inf Model 59:3370–3388

15. Sarker IH (2021) Machine learning: Algorithms, real-world applications and research directions. SN Comput Sci 2:1–21

16. Xu P (2019) Review on studies of machine learning algorithms. J Phys Conf Ser. https://doi.org/10.1088/1742-6596/1187/5/052103

17. Mohammed M, Khan MB, Bashie EBM (2016) Machine learning: Algorithms and applications. Mach Learn Algorithms Appl. https://doi.org/10.1201/9781315371658

18. Sarker IH, Kayes ASM, Badsha S, Alqahtani H, Watters P, Ng A (2020) Cybersecurity data science: An overview from machine learning perspective. J Big Data. https://doi.org/10.1186/s40537-020-00318-5

19. Arden NS, Fisher AC, Tyner K, Yu LX, Lee SL, Kopcha M (2021) Industry 4.0 for pharmaceutical manufacturing: Preparing for the smart factories of the future. Int J Pharm. https://doi.org/10.1016/j.ijpharm.2021.120554

20. Pisner DA, Schnyer DM (2019) Support vector machine. In: Mechelli A, Vieira S (eds). Machine Learning: Methods and Applications to Brain Disorders. Academic Press, pp 101–121

21. Usama M, Qadir J, Raza A, Arif H, Yau KLA, Elkhatib Y, Hussain A, Al-Fuqaha A (2019) Unsupervised machine learning for networking: Techniques, applications and research challenges. IEEE Access 7:65579–65615

22. Yu SS, Chu SW, Wang CM, Chan YK, Chang TC (2018) Two improved k-means algorithms. Appl Soft Comput J 68:747–755

23. Bourquin J, Schmidli H, Van Hoogevest P, Leuenberger H (1998) Comparison of artificial neural networks (ANN) with classical modelling techniques using different experimental designs and data from a galenical study on a solid dosage form. Eur J Pharm Sci 6:287–300

24. Kesavan JG, Peck GE (1996) Pharmaceutical granulation and tablet formulation using neural networks. Pharm Dev Technol 1:391–404

25. Turkoglu M, Aydin I, Murray M, Sakr A (1999) Modeling of a roller-compaction process using neural networks and genetic algorithms. Eur J Pharm Biopharm 48:239–245

26. Bannigan P, Flynn J, Hudson SP (2020) The impact of endogenous gastrointestinal molecules on the dissolution and precipitation of orally delivered hydrophobic APIs. Expert Opin Drug Deliv 17:677–688

27. Van Hoogevest P, Liu X, Fahr A (2011) Drug delivery strategies for poorly water-soluble drugs: The industrial perspective. Expert Opin Drug Deliv 8:1481–1500

28. Damiati SA, Martini LG, Smith NW, Lawrence JM, Barlow DJ (2017) Application of machine learning in prediction of hydrotrope-enhanced solubilisation of indomethacin. Int J Pharm 530:99–106

29. Yang Y, Ye Z, Su Y, Zhao Q, Li X, Ouyang D (2019) Deep learning for *in vitro* prediction of pharmaceutical formulations. Acta Pharm Sin B 9:177–185

30. Dowell JA, Hussain A, Devane J, Young D (1999) Artificial neural networks applied to the *in vitro–in vivo* correlation of an extended-release formulation: Initial trials and experience. J Pharm Sci 88:154–160

31. Mendyk A, Tuszyński PK, Polak S, Jachowicz R (2013) Generalized *in vitro–in vivo* relationship (IVIVR) model based on artificial neural networks. Drug Des Devel Ther 7:223–232

32. He Y, Ye Z, Liu X, Wei Z, Qiu F, Li HF, Zheng Y, Ouyang D (2020) Can machine learning predict drug nanocrystals? J Control Release 322:274–285

33. Rantanen J, Khinast J (2015) The future of pharmaceutical manufacturing sciences. J Pharm Sci 104:3612–3638

34. Paul D, Sanap G, Shenoy S, Kalyane D, Kalia K, Tekade RK (2021) Artificial intelligence in drug discovery and development. Drug Discov Today 26:80–93

35. Simões MF, Silva G, Pinto AC, Fonseca M, Silva NE, Pinto RMA, Simões S (2020) Artificial neural networks applied to quality-by-design: From formulation development to clinical outcome. Eur J Pharm Biopharm 152:282–295

36. Huang J, Kaul G, Cai C, Chatlapalli R, Hernandez-Abad P, Ghosh K, Nagi A (2009) Quality by design case study: An integrated multivariate approach to drug product and process development. Int J Pharm 382:23–32

37. Sokolović N, Ajanović M, Badić S, Banjanin M, Brkan M, Čusto N, Stanić B, Sirbubalo M, Tucak A, Vranić E (2020) Predicting the outcome of granulation and tableting processes using different artificial intelligence methods. In: Badnjevic, A, Škrbić, R, Gurbeta Pokvić, L. (eds). CMBEBIH 2019. IFMBE Proceedings, vol 73. Springer. pp 499–504

38. Landin M (2017) Artificial intelligence tools for scaling up of high shear wet granulation process. J Pharm Sci 106:273–277

39. Roggo Y, Jelsch M, Heger P, Ensslin S, Krumme M (2020) Deep learning for continuous manufacturing of pharmaceutical solid dosage form. Eur J Pharm Biopharm 153:95–105

40. Gams M, Horvat M, Ožek M, Luštrek M, Gradišek A (2014) Integrating artificial and human intelligence into tablet production process. AAPS PharmSciTech 15:1447–1453

41. Grof Z, Štěpánek F (2021) Artificial intelligence based design of 3D-printed tablets for personalised medicine. Comput Chem Eng 154:107492. https://doi.org/10.1016/j.compchemeng.2021.107492

42. Mariano G, Farthing RJ, Lale-Farjat SLM, Bergeron JRC (2020) Structural characterization of SARS-CoV-2: Where we are, and where we need to be. Front Mol Biosci 17:605236. https://doi.org/10.3389/fmolb.2020.605236

43. (2020) Artificial intelligence in oncology drug discovery and development. Artif Intell Oncol Drug Discov Dev. https://doi.org/10.5772/intechopen.88376

44. Egorov E, Pieters C, Korach-Rechtman H, Shklover J, Schroeder A (2021) Robotics, microfluidics, nanotechnology and AI in the synthesis and evaluation of liposomes and polymeric drug delivery systems. Drug Deliv Transl Res 11:345–352

45. Shalaby KS, Soliman ME, Casettari L, Bonacucina G, Cespi M, Palmieri GF, Sammour OA, El Shamy AA (2014) Determination of factors controlling the particle size and entrapment efficiency of noscapine in PEG/PLA nanoparticles using artificial neural networks. Int J Nanomedicine 9:4953–4964

46. Khong J, Wang P, Gan TRX, Ng J, Anh L, Blasiak TT, Kee A, Ho TD (2019) The role of artificial intelligence in scaling nanomedicine toward broad clinical impact. In: Nanoparticles for Biomedical Applications: Fundamental Concepts, Biological Interactions and Clinical Applications. Elsevier, pp 385–407

47. Suhail M, Khan A, Rahim MA, Naeem A, Fahad M, Badshah SF, Jabar A, Janakiraman AK. (2022) Micro and nanorobot-based drug delivery: An overview. J Drug Target 30(4):349–358. https://doi.org/10.1080/1061186X.2021.1999962.

48. Pokharkar D, Shetty T, Thorat M, Yadav N (2021) Nanorobotics: A booming trend in pharmaceutical industry. World J Pharm Res 10:307–323

49. Singh AV, Chandrasekar V, Janapareddy P, et al (2021) Emerging application of nanorobotics and artificial intelligence to cross the BBB: Advances in design, controlled maneuvering, and targeting of the barriers. ACS Chem Neurosci 12:1835–1853

50. Jiang J, Ma X, Ouyang D, Williams RO (2022) Emerging artificial intelligence (AI) technologies used in the development of solid dosage forms. Pharmaceutics 14:2257

51. Solomun L, Ibric S, Pejanovic V, Djuris J, Jockovic J, Stankovic P, Vujic Z (2012) *In silico* methods in stability testing of hydrocortisone, powder for injections: Multiple regression analysis versus dynamic neural network. Hem Ind 66:647–657

52. Ajdarić J, Ibrić S, Pavlović A, Ignjatović L, Ivković B (2021) Prediction of drug stability using deep learning approach: Case study of esomeprazole 40 mg freeze-dried powder for solution. Pharmaceutics 13:829. https://doi.org/10.3390/pharmaceutics13060829

53. Gu Y, Zalkikar A, Liu M, Kelly L, Hall A, Daly K, Ward T (2021) Predicting medication adherence using ensemble learning and deep learning models with large scale healthcare data. Sci Rep 11:18961. https://doi.org/10.1038/s41598-021-98387-w

54. Wang S, Di J, Wang D, Dai X, Hua Y, Gao X, Zheng A, Gao J (2022) State-of-the-art review of artificial neural networks to predict, characterize and optimize pharmaceutical formulation. Pharmaceutics 14:183. https://doi.org/10.3390/pharmaceutics14010183

55. Hyland SL, Faltys M, Hüser M, et al (2020) Early prediction of circulatory failure in the intensive care unit using machine learning. Nat Med 26:364–373

56. Moreau JT, Hankinson TC, Baillet S, Dudley RWR (2020) Individual-patient prediction of meningioma malignancy and survival using the surveillance, epidemiology, and end results database. NPJ Digit Med 3:12. https://doi.org/10.1038/s41746-020-0219-5

57. Yu SC, Gupta A, Betthauser KD, Lyons PG, Lai AM, Kollef MH, Payne PRO, Michelson AP (2022) Sepsis prediction for the general ward setting. Front Digit Heal 4:1–8

58. Karwasra R, Khanna K, Singh S, Ahmad S, Verma S (2022) The incipient role of computational intelligence in oncology: Drug designing, discovery, and development. In: Computational Intelligence in Oncology. Springer, pp 369–384

59. Amgen. Amgen and Generate Biomedicines Announce Multi-Target, Multi-Modality Research Collaboration Agreement [Internet]. 2022. Available from: https://www.amgen.com/newsroom/press-releases/2022/01/amgen-and-generate-biomedicines-announce-multitarget-multimodality-research-collaboration-agreement (accessed Nov 29, 2022)

60. Owkin. Owkin becomes 'unicorn' with $180m investment from Sanofi and four new collaborative projects [Internet]. 2021. Available from: https://owkin.com/publications-and-news/press-releases/owkin-becomes-unicorn-with-180m-investment-from-sanofi-and-four-new-collaborative-projects   (accessed Nov 29, 2022)
61. Owkin. Owkin announces multi-year clinical data science strategic collaboration with Bristol Myers Squibb [Internet]. 2022. Available from: https://owkin.com/publications-and-news/press-releases/owkin-announces-multi-year-clinical-data-science-strategic-collaboration-with-bristol-myers-squibb (accessed Nov 29, 2022)
62. https://www.fda.gov/files/medical%20devices/published/US-FDA-Artificial-Intelligence-and-Machine-Learning-Discussion-Paper.pdf

8 Pharmaceutical Drugs, Biologics, Gene Therapies, and Medical Device Regulation

Kirubakaran Narayanan
SRM College of Pharmacy, Kattankulathur, Tamil Nadu, India

K. Reeta Vijaya Rani
Surya School of Pharmacy, Vikravandi, Tamil Nadu, India

Umashankar Marakanam Srinivasan
SRM College of Pharmacy, Kattankulathur, Tamil Nadu, India

Kiruba Mohandoss
Sri Ramachandra Institute of Higher Education and
Research, Chennai, Tamil Nadu, India

Yoga Senbagapandian Rajamani
University of Maryland, College Park, Maryland, USA

Hanan Fahad Alharbi
Princess Nourah bint Abdul Rahman University, Riyadh, Saudi Arabia

S Anbazhagan
Surya School of Pharmacy, Vikravandi, Tamil Nadu, India

8.1 INTRODUCTION

Pharmaceutical drugs, biologics, gene therapies, and medical devices are very important in health-care sector. "Pharmaceuticals" and "pharmaceutical drugs" encompass a different kind of medicines used to prevent, diagnose, treat, or cure diseases in the society. The pharmaceutical drugs section of the National Alcohol and Drug Knowledgebase (NADK) addresses a subsection of these medicines like Unscheduled Medicines, Schedule 2—Pharmacy medicines, Schedule 3—Pharmacist-only medicines, Schedule 4—Prescription-only medicines, etc. (1).

A biological drug (biologics) is a product that is derived from directly living organisms or which may contain components of living organisms. Biological drugs include a wide variety of products derived from human, animal, or microorganisms by using biotech technology. The variety of biological drugs include vaccines, blood, blood components, cells, allergens, genes, tissues, recombinant proteins, etc. Biological products may contain proteins that can control the action of other proteins and cellular processes, genes that control production of vital proteins, modified human hormones, or cells that produce substances that suppress or activate components of the immunological

DOI: 10.1201/9781003343981-8

system. Biological drugs are also called biological response modifiers due to their ability to change the manner of operation of natural biological intracellular and cellular actions. Biological drugs are used for treatment of numerous diseases and conditions, for example, Crohn's disease, ulcerative colitis, rheumatoid arthritis, and other autoimmune diseases (2).

Gene therapy is a biomedical field which focuses on the genetic modification of cells to produce a therapeutic benefit (3) or the treatment of disease by repairing or reconstructing defective genetic material (4). The first attempt made at modifying human DNA was performed by Martin Cline in 1980, but the first successful nuclear gene transfer in humans, approved by the National Institutes of Health, was achieved in 1989 (5). French Anderson *performed* the first therapeutic use of gene transfer as well as the first direct insertion of human DNA into the nuclear genome in 1990 on trial basis. It is thought to be able to cure many genetic disorders or treat them over time. The concept of gene therapy is to fix a genetic problem at its level of source conditions. If, for instance, a mutation in a certain gene causes the production of a dysfunctional protein which result in an inherited disease, gene therapy could be used to deliver a copy of this gene that does not contain the deleterious mutation and thereby produces a functional protein. This strategy is called gene replacement therapy and this principle is used to treat the inherited retinal diseases (6, 7). The concept of gene replacement therapy is mostly suitable for recessive diseases (8).

8.2   CLASSIFICATION OF PHARMACEUTICAL DRUGS

8.2.1   Unscheduled Medicines

These medicines are available for general sale in supermarkets, grocery stores, health food stores, and pharmacies, often with labels about safe use (e.g., for non-prescription pain relief medicines). In the NADK, these drugs include formulations like aspirin, paracetamol, and ibuprofen.

8.2.2   Schedule 2: Pharmacy Medicines

These medicines are available on open shelves only at pharmacies, but a pharmacist or pharmacy assistant must be available for advice if required with respect to dose and dosage instructions (e.g., larger packets of non-prescription pain relief medicines). In the NADK, these drugs include formulations of aspirin, paracetamol, and ibuprofen.

8.2.3   Schedule 3: Pharmacist-only Medicines

These medicines are only available from behind the counter at a pharmacy. No prescription is required, but a pharmacist must be consulted before they are dispensed. In the NADK, these drugs include formulations like analgesic ibuprofen and paracetamol.

8.2.4   Schedule 4: Prescription-only Medicines

These medicines must be prescribed by an authorized healthcare professional and may be supplied to hospitals or purchased from a pharmacy with a prescription. In the NADK, these include tramadol, codeine-containing medicines, and most benzodiazepines kind of drugs.

8.2.5   Schedule 8: Controlled Drugs

These medicines must be prescribed by an authorized healthcare professional, who may need a permit to prescribe them. Drugs included in the NADK are opioids (excluding formulations used for opioid substitution therapy) such as buprenorphine, codeine, fentanyl, hydromorphone, methadone, morphine, oxycodone and tapentadol, and the benzodiazepines flunitrazepam and alprazolam.

8.3 CLASSIFICATION OF GENE THERAPY

8.3.1 SOMATIC

In somatic cell gene therapy (SCGT), the therapeutic genes are transferred into any cell other than a gamete, germ cell, gametocyte, or undifferentiated stem cell. Any such modifications affect the individual patient only and are not inherited by offspring. Somatic gene therapy represents mainstream basic and clinical research, in which therapeutic DNA (either integrated in the genome or as an external episome or plasmid) is used to treat disease (9). In the United States, over 600 clinical trials utilized SCGT. The application and research of SCGT focus on severe genetic disorders, which include immunodeficiencies, hemophilia, thalassemia, and cystic fibrosis. Mainly single-gene disorders are cured by somatic cell therapy. The complete correction of a genetic disorder or the replacement of multiple genes is not yet possible (10).

8.3.2 GERMLINE

In germline gene therapy (GGT), germ cells (sperm or egg cells) are modified by the introduction of functional genes into their genomes. Modifying a germ cell causes all the organism's cells to contain the modified genes. The change is therefore heritable and passed on to later generations. Some of the countries like Australia, Canada, Germany, Israel, Switzerland, and the Netherlands (11) prohibit GGT application in human beings for technical and ethical reasons, including insufficient knowledge about possible risks to future generations and higher risks versus SCGT (12). The United States has no federal controls specifically addressing human genetic modification (beyond Food and Drug Administration [FDA] regulations for therapies in general) (13–15).

According to the European definition, a medical device is "any instrument, apparatus, appliance, material or other article, whether used alone or in combination, including the software necessary for its proper application intended by the manufacturer to be used for human beings for the purpose of:

- Diagnosis, prevention, monitoring, treatment or alleviation of disease;
- Diagnosis, monitoring, treatment, alleviation of or compensation for an injury or disability;
- Investigation, replacement or modification of the anatomy or of a physiological process;
- Control of conception;

and which does not achieve its principal intended action in or on the human body by pharmacological, immunological or metabolic means, but which may be assisted in its function by such means" (16). Medical devices are usually divided into subgroups according to the region or country with main focus on the basis of risk (17). Today, there are more than 8,000 generic medical device groups, among those some of the devices contain drugs (18). This increases the need for better regulatory frameworks to ensure that products entering the market are safe and efficient for their consumers as well as respective category of the patients. One of the major challenges and issues for companies developing and producing medical devices is to meet the criteria or the regulatory requirements for implementation of the same (19).

8.4 CLASSIFICATION OF MEDICAL DEVICES

Medical devices are usually divided into different classes. Some countries have separate classification systems for general medical devices, active medical devices for implantation, and *in vitro* diagnostic devices, while other countries have same classification systems for these products. All classification systems are based on risk criteria. Classification of medical devices is necessary to apply correct regulations and quality systems. In the United States, medical devices are classified as class I (General Controls), II (Special Controls), or III (Pre-market Approval) devices where class III devices represent the highest risk and require more control (20).

In the European Union (EU), general medical devices are classified as class I, class I sterile, class I measuring, class IIa, class IIb, or class III where class III devices represent the highest risk. Active implantable medical devices are not classified, and *in vitro* diagnostic devices have their own classification system.

8.4.1 Regulations of Medical Device

Manufacturers of medical devices need to follow the regulatory guidelines in the country where the product is sold. This constitutes a great problem for manufacturers, especially for companies selling their products in several countries. Competent authorities worldwide have begun to realize the problem and collaborate to harmonize the regulations and bring into one platform worldwide regulatory guidelines. The Global Harmonization Task Force (GHTF) is a group of representatives from regulatory authorities in the United States, EU, Japan, Australia, and Canada that work to harmonize the regulations for medical devices and improve the safety, effectiveness, and quality of the device's products. The group has developed guidelines for pre-market evaluation, post-market surveillance, quality systems, auditing, clinical safety/performance, etc. Many countries have begun to adopt these guidelines or follow the FDA regulations or the European Medicines Agency (EMEA) regulations. Medical device requirements are basically the same in most countries but there are difference in the implementation process (21).

In the United States, the FDA regulates food, drugs, medical devices, biologics, cosmetics, and radiation-emitting products in the United States. FDA's Center for Devices and Radiological Health (CDRH) is responsible for regulating manufacturers of medical devices as per the framed regulatory guidelines. Medical devices are regulated under the Federal Food Drug & Cosmetic Act (FD&C Act) Part 800-1299 guidelines. Manufacturers importing medical devices into the United States must designate a US agent, register the establishment, list the device, manufacture according to the quality system requirements, and file a Premarket Notification 510 (k) or a Premarket Approval. A post-marketing surveillance system is required (21 CFR Part 803). Medical devices are divided into Class I, Class II, and Class III, where Class I devices represent the lowest risk and Class III devices represent the highest risk. Most Class I devices and some Class II devices are exempt from a Premarket Notification 510(k). Class II devices generally require a 510(k), while Class III devices require a Premarket Approval. Devices shall be given a device product code consisting of two numbers and three letters describing what type of device it is. Regulations for establishment registration and medical device are listed in 21 CFR 807 guidelines. The establishment registration shall be renewed once a year time frame and the device listing updated once a year duration of the time. Good Manufacturing Practice (GMP) shall be applied according to the requirements of 21 CFR Part 820. Some devices are exempted from GMP requirements like Class I segmented medical devices. The international standards for risk management ISO 14971 and biocompatibility ISO 10993 are recognized as well as accepted in the license applications of medical devices (22).

The EMEA is a decentralized body of the EU whose responsibility is to protect as well as safeguard the human and animal health through the evaluation and supervision of medical products for the application of human or animal usage. This information and guidelines are listed in EMEA homepage. Medical devices are subject to Directive 93/42/EEG and must be CE-marked before entering any country in the EU region. Active implantable medical devices should follow the guidelines of Directive 90/385/EEG (23). Manufacturers of drugs and medical devices who want to sell their products to a country in the EU only need to submit one single marketing authorization application to the EMEA for the approval process. The documentation shall be written in English, French, or German language for the submission process. A manufacturer that does not have a registered place of business in the EU shall designate a single authorized representative in the EU region where they want to sell the products. Medical devices are divided into class I, class IIa, class IIb, and class III, where class I also have the subclasses—sterile and measuring. The devices shall have a GMDN code. All medical devices exempt class I devices require the involvement of a Notified Body. The accessories of medical devices are also treated as medical devices. Medical devices must meet the essential requirements in Annex I of Directive 93/42/EEG and guidelines. Standards are used to meet and demonstrate compliance with the essential requirements to ensure safety. Manufacturers of medical devices must adhere the quality system principles. ISO 13485 is normally used for medical devices. Clinical trials are applicable for active implantable devices, class III devices, and invasive devices for long-term use of class IIa and IIb. Instructions for use are not necessary for class I and IIa devices if they can be used safely without them. The registration of a product is valid for five years in the EU region (24).

Most countries have similar requirements for registration of medical devices and are striving to harmonize with the GHTF guidelines. Classification of medical devices is usually done in accordance with the EU system, FDA system, GHTF guidelines, or by catalog. UMDNS codes or GMDN are used for the nomenclature where GMDN seems to be the most common variant. Main requirements are a local representative, a Certificate of Free Sale from the country of origin, import license from the competent authority in the import country, and registration of the company and the product. Quality management systems and risk management systems are in most countries required, except for medical devices class I. Certificates of ISO 13485 and ISO 14971 are required or recommended to ensure the Quality Management systems.

8.5   SOME OF THE COUNTRIES AND THEIR REGULATORY BODIES

8.5.1   Australia

The Therapeutic Goods Administration (TGA) is the competent authority for medical devices in Australia (Figure 8.1). TGA is a unit of the Australian Government Department of Health and Ageing and is responsible for administering the provisions of the legislation under the Therapeutic Goods Act 1989 (the Act) (Figure 8.3). This Act covers both medical devices and AIMDs. The organization is divided into several parts where the Office of Devices, Blood and Tissues is responsible for medical devices.

Medical devices must be registered in the database Australian Register of Therapeutic Goods (ARTG) before entering the Australian market. Sponsors are recommended to use the Devices Electronic Application Lodgement (DEAL) system for the applications (25, 26).

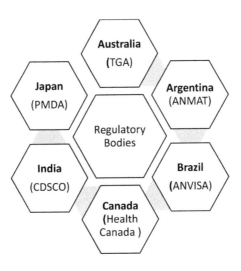

FIGURE 8.1   Different Countries (AABCIJ) and their Regulatory agencies.

8.5.2 ARGENTINA

Medical devices are regulated by the National Administration of Drugs, Foodstuffs and Medical Technology (ANMAT) under the Ministry of Health. Medical products in Argentina and importers of medical devices must be registered with ANMAT (Figure 8.1). The importer is responsible for registration of medical devices as per the ANMAT. Technical information shall be submitted with the registration including documents that are legalized by the Argentine Consulate or Embassy in the product's country of origin. Additional voluntary technical regulations are issued by the Standards Institute of Argentina (IRAM) (14). According to resolution 3802/2004, a medical device produced in or imported to Argentina must show conformity with Mercosur Technical Regulations for Registration of Medical Products (Figure 8.3). The resolution gives the definition of a medical device, describes labeling conditions, and recognizes established risk assessment categories and the nomenclature of medical device. The Argentine ANMAT 10 disposition 2318/2002 defines a medical device and describes the registering process, the classification rules, the required technical documentation, etc. Argentina has followed the Emergency Care Research Institute (ECRI) nomenclature called the Universal Medical Device Nomenclature System (UMDNS). The product must be classified according to the MERCOSUR system (27). All medical devices shall meet the essential needs in disposition 4306/99. GMP for medical devices shall be applied and certified according to the disposition 191/9 guidelines. Recognized standards are ISO/TR 16142, ISO 17025, ISO 14971, and IEC 60601-61610 for the medical device products (28).

8.5.3 BRAZIL

The National Health Surveillance Agency (ANVISA) or in Portuguese Agencia Nacional de Vigilancia Sanitaria (ANVISA) is the competent authority for medical devices in Brazil (Figure 8.1). All medical devices, diagnostic kits, immune-biological products, and sanitation products must be registered with ANVISA before getting out on the Brazilian market. Medical devices are regulated by Law No. 6360 of 1976, decree 74.094/97. Resolution RDC-185 of October 22, 2001, is the main resolution for medical devices (Figure 8.3). This resolution describes the required documents for registering a product and it contains a registration protocol. Resolution RDC No. 206 of November 2006 describes the requirements for registering *in vitro* diagnostic devices for licensing purposes. A certificate of Brazilian GMP (in Portuguese Boas Practicas) is required, and a copy of the certificate must be submitted with the registration and application process. Active medical devices under IEC 60601-1 should be certified by an INMETRO accredited test agency. Risk management is mandatory for all implantable devices, intrauterine devices, and plastic bags for blood. Brazil has adhered to the risk management standard ISO 14971 as a national standard. Requirements for risk factors are found in the essential principles for medical device applications (29, 30). Clinical trials need new products and products with innovative technology aspects (31).

8.5.4 CANADA

Health Canada, under the authority of the Food and Drugs Act, regulates the sale of drugs and medical devices in Canada. Health Canada is divided into two parts: Health Products and Food, and Therapeutic Products Directorate (Figure 8.1). Medical Devices Bureau is under Therapeutic Products Directorate which is divided into Device Evaluation, Licensing Services, and Research and Surveillance (32). Medical devices in Canada are subject to the Medical Devices Regulations (referred to as the Regulations) under the Food and Drugs Act (Figure 8.3) (33). Canada has adopted ISO 13485:2003 as a Canadian National Standard and labeled it CAN/CSA-ISO 13485:2003.

8.5.5 INDIA

The Department of Health under India's Ministry of Health and Family Welfare is responsible for the jurisdiction over the regulation of medical devices. The Central Drug Standard Control

Organization (CDSCO) in the Ministry of Health is primarily responsible for regulation of drugs but also medical devices, diagnostic devices, and cosmetics. Pharmaceuticals and medical devices defined as drugs are regulated under the Drug and Cosmetics Act 1940 (Figure 8.3) and the Drugs and Cosmetic Rules 1945 and must be registered before they can be sold in India (Figure 8.1). Medical devices defined as drugs require a registration certificate and an import license before being sold on the Indian market. The Medical Devices Regulation Bill 2006 is introduced to monitor regulations of medical device in India (34, 35).

8.5.6 Japan

The Ministry of Health, Labor, and Welfare (MHLW) is responsible for food, medical care, labor policy and labor standards, and social welfare. The Pharmaceutical and Food Bureau within the ministry is responsible for pharmaceutical and medical device regulatory policymaking. An instance called the Pharmaceuticals and Medical Devices Agency (PMDA) is responsible for the registration of medical devices (Figure 8.1). In 2005, a new law came into effect which is harmonized with international requirements. The law is called the New Pharmaceutical Affairs Law (PAL) (Figure 8.3) (36). The main difference with international requirements is that Japan has additional requirements for buildings and facilities of manufacturing sites for medical devices (37). A manufacturer must adhere to the principles of Market Authorization Holder (MAH) system. With this system, the manufacturer is only responsible for production and the MAH is responsible for the release of the product to the market. Japan also has requirements on quality systems and risk management. For manufacturers of medical devices, certificates of Japanese GMP and Good Vigilance Practice (GVP) are Mandatory.

8.5.7 Mexico

Medical devices are regulated by the Secretariat of Health (Secretaría de Salud) (38). According to Article 262o of the Mexican General Health Law (Figure 8.2), the medical devices must be registered with the Secretariat of Health before entering the marketing (Figure 8.3). The Federal Commission for Protection of Sanitary Risks (COFEPRIS) is in charge of registering any healthcare product (39). It permits licenses authorization and regulation. It is responsible for maintaining the standards, sanitary risk management, auditing vigilance, and regulation compliance with law enforcement (40). Medical devices (and *in vitro* diagnostic devices) intended to be sold and used by consumers on the Mexican market must adhere to the mandatory labeling standard NOM-137-SSA1-1995.

8.5.8 Russia

The Federal Service for Control over Healthcare and Social Development (Roszdravnadzor) (Figure 8.2) is the competent authority in Russia for registration of medical devices. Foreign manufacturers work through the Department of Registration of Foreign Medical Equipment and Devices. In June 2000, a new instruction, Instruction No. 237 (Figure 8.3), on registration procedures for foreign-made medical equipment and devices was introduced. Foreign manufacturers of medical devices must register their product with the competent authority and obtain a GOST-R quality and safety certification. The product must be defined as a medical device and classified according to the OKP system for registration and TNVED for importation (41, 42).

8.5.9 South Korea

The Ministry of Health and Welfare (MHW) is the competent authority responsible for the import of medical devices. Korea Food and Drug Administration (KFDA) is responsible for the regulatory body (Figure 8.2). This agency regulates all medical devices under Korea's Medical Device Act (Figure 8.3)

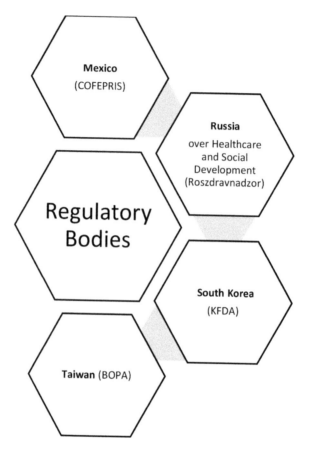

FIGURE 8.2   Different Countries (MRST) and their Regulatory agencies.

which was passed by Korea's National Assembly in 2003, implemented in 2004, and fully enforced on May 30, 2007. All medical devices require premarket registration from KFDA and adhere to Korean Good Manufacturing Practice (KGMP) before they can be imported into Korea or manufactured in Korea (43). All medical devices must meet KGMP according to the Medical Device Act. Foreign manufacturers can fulfill good manufacturing requirements by following ISO 13485 or comply with the US quality requirements (44).

8.5.10   Taiwan

Medical devices are regulated by the Bureau of Pharmaceutical Affairs (BOPA) under the Department of Health (DOH) (Figure 8.2). The Pharmaceutical Affairs Law (Figure 8.3) sets the requirements for medical devices. On April 12, 2006, a revised version of Guidelines for Registration of Medical Devices was created. *In vitro* diagnostic devices are regulated as medical devices category (45–47).

8.6   REGULATIONS OF GENE THERAPY PRODUCTS (GTPs)

8.6.1   EMA (EU)

In EU, the gene therapy medicinal product (GTMP) is defined as a biological medicinal product (excluding vaccines) that (a) contains an active substance which contains or consists of a recombinant nucleic acid used in or administered to human beings with a view to regulating, repairing, replacing, adding or deleting a genetic sequence and; (b) its therapeutic, prophylactic, or diagnostic

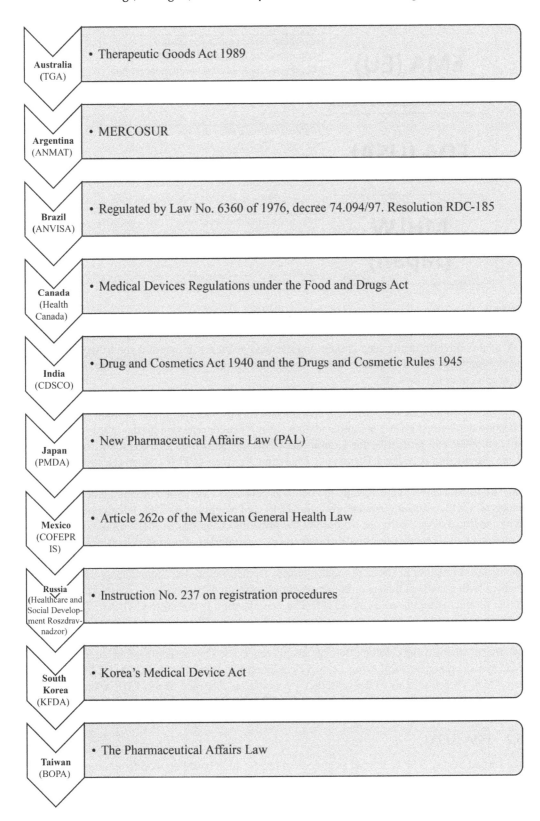

FIGURE 8.3 Different Countries with Regulatory Act and Law's.

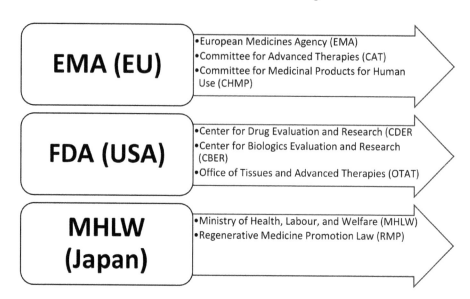

FIGURE 8.4 Regulations of gene therapy products in EU-USA-JAPAN.

effect relates directly to the recombinant nucleic acid sequence it contains, or to the product of genetic expression of this sequence. Gene therapy medicinal products shall not include vaccines against infectious diseases (48).

The European Medicines Agency (EMA) is the centralized regulatory body of the EU. Within the EMA, the Committee for Advanced Therapies (CAT) provides an expert opinion on the drug's application dossier and gives a recommendation for approval or rejection for the process. This decision is reviewed and ratified by the Committee for Medicinal Products for Human Use (CHMP). GTs are classified as Advanced Therapeutic Medicinal Products (ATMP) which are governed under the ATMP Regulation (Directive 2001/83/EC, as amended by Regulation (EC) No. 1394/2007). ATMP Regulation covers gene therapy medicinal products (GTMP), somatic cell therapy medicinal products (CTMP), tissue engineered products (TEP), and combined ATMPs, which are a combination of a medical device with one or more of the previous categories. ATMPs should comply with EU pharmaceutical regulations and therefore receive premarket approval. A non-pharmaceutical product meets the following criteria: "It is not substantially manipulated; cells are used for the same essential function in the donor and recipient (sometimes called 'homologous use'); it is not combined with a medical device or an active implantable medical device." If these criteria are not met, the product is regulated as an ATMP. The classification is tasked to CAT, which provides a non-binding recommendation as to whether a specific product should be considered an ATMP. As with 351/361 Product designation, any GTP would be considered a GTMP, and therefore an ATMP, due to the mechanism and cellular effect of gene therapeutics. The regulation of a Combined ATMP is done by its components, following a categorical hierarchy defined by the EMA: GTMP over TEP over CTMP. Therefore, an *ex vivo* gene therapy would be classified as a GTMP, even though the final therapeutic product would be cell-based like CTMPs (Figure 8.4) (49).

8.6.2   FDA (USA)

In the United States, GTP is defined as a medical intervention based on modification of the genetic material of living cells. Cells may be modified *ex vivo* for subsequent administration to humans or may be altered *in vivo* by gene therapy given directly to the subject. When the genetic manipulation is performed *ex vivo* on cells which are then administered to the patient, this is also a form of somatic cell therapy. The genetic manipulation may be intended to have a therapeutic or prophylactic effect or may provide a way of marking cells for later identification. Recombinant DNA materials used to transfer genetic material for such therapy are considered components of gene therapy and as such are subject to regulatory oversight (50).

The USFDA has jurisdiction over a variety of products, including food, tobacco, vaccines, and therapeutics in the United States. Within the FDA, therapeutics are regulated under either the Center for Drug Evaluation and Research (CDER) (Figure 8.4) or the Center for Biologics Evaluation and Research (CBER). CBER is responsible for the regulation of human gene therapy–related products (51). Within CBER, oversight of GTPs falls to the Office of Tissues and Advanced Therapies (OTAT), which before restructuring in October 2016 was known as the Office of Cellular, Tissue, and Gene Therapies (OCTGT) (52). In addition to regulatory review, OTAT also releases regulatory policy and guidance documents, many of which will be referenced in this publication (53, 54).

8.6.3 MHLW (Japan)

In Japan, the term "Regenerative Medicinal Products" (as "SAISEI-IRYOUTOU-SEIHIN" in Japanese) used in this act refers to the articles (excluding quasi-drugs and cosmetics) specified in the following items which are specified by the cabinet order. The articles which are intended to be used in the treatment of disease in humans or animals are transgened to express in human or animal cells (55).

Japan's regulatory agency, MHLW (Figure 8.4) has a special regulatory framework for "regenerative medicinal products" under which GTPs are categorized and regulated (not to be confused with "regenerative medicine," a separate category defined in the Japanese regulation solely for academic or clinical research products where market approval will not be sought). The Japanese government recently passed a set of legislations to promote the development of Japan as an international hub of medical research and therapeutics development, possibly due in part to the important role of Japanese scientists in the discovery of induced pluripotent stem cell (iPSC). Of specific relevance to regenerative medicines, and therefore GTPs, are the Regenerative Medicine Promotion Law (RMP, May 2013), the Act of Safety of Regenerative Medicine (RM Act, November 2013), and the Act on Pharmaceuticals and Medical Devices (PMD Act, November 2013). This legislation and statements from the MHLW have demonstrated their interest in promoting iPSC-based therapeutics development (56). The RMP guarantees broad governmental protection and support of regenerative medicines at all stages and the RM Act gives more granular regulation on the clinical development of regenerative medicines, regulating logistical aspects to regenerative medicine development such as risk classification and processing facility requirements (57–59).

REFERENCES

1. Nicholas R, Lee N, Roche A et al, 2011. Pharmaceutical Drug Misuse Problems in Australia: Complex Issues, Balance Responses. National Centre for Education and Training on Addiction (NCETA), Flinders University.
2. Chavda VP et al, 2019. Nanotherapeutics and Nanobiotechnology. Applications of Targeted Nano Drugs and Delivery Systems. Elsevier. Pages 1–13.
3. Kaji Eugene H et al, 2001. Gene and Stem Cell Therapies. JAMA. 285(5), Pages 545–550.
4. Ermak G et al, 2015. Emerging Medical Technologies. World Scientific Publishing Company. Pages 1–152.
5. Rosenberg SA, Aebersold P, Cornetta K et al, 1990. Gene Transfer into Humans: Immunotherapy of Patients with Advanced Melanoma, Using Tumor-Infiltrating Lymphocytes Modified by Retroviral Gene Transduction. The New England Journal of Medicine. 323(9), Pages 570–578.
6. Maguire AM, Simonelli F, Pierce EA et al, 2008. Safety and Efficacy of Gene Transfer for Leber's Congenital Amaurosis. The New England Journal of Medicine. 358(21), Pages 2240–2248.
7. MacLaren RE, Groppe M, Barnard AR et al, 2014. Retinal Gene Therapy in Patients with Choroideremia: Initial Findings from a Phase 1/2 Clinical Trial. Lancet. 383(9923), Pages 1129–1137.
8. Bak RO, Gomez-Ospina N, Porteus MH et al, 2018. Gene Editing on Center Stage. Trends in Genetics. 34(8), Pages 600–611.

9. Williams DA, Orkin SH et al, 1986. Somatic Gene Therapy: Current Status and Future Prospects. The Journal of Clinical Investigation. 77(4), Pages 1053–1056.

10. Mavilio F, Ferrari G et al, 2008. Genetic Modification of Somatic Stem Cells: The Progress, Problems, and Prospects of a New Therapeutic Technology. EMBO Reports. 9(1), Pages S64–S69.

11. "International Law". The Genetics and Public Policy Center, Johns Hopkins University Berman Institute of Bioethics. 2010.

12. Strachnan T, Read AP et al, 2004. Human Molecular Genetics (3rd ed.). Garland Publishing. Page 616.

13. Hanna K et al, 2006. "Germline Gene Transfer". National Human Genome Research Institute.

14. "Human Cloning and Genetic Modification", 2013. Association of Reproductive Health Officials.

15. "Gene Therapy", 2014. ama-assn.org.

16. European Commission, Directive 93/42/EEC, published June 14, 1993, and valid till December 31, 2008, at http://eur-lex.europa.eu.

17. Landvall P et al, 2007. Medicintekniska produkter – Vägledning till CEmärkning, SIS Förlag AB, Stockholm.

18. World Health Organization, 2007. Sixtieth World Health Assembly, Provisional Agenda Item 12.19.

19. Iglesias-Lopez C, Agustí A, Obach M et al, 2019. Regulatory Framework for Advanced Therapy Medicinal Products in Europe and United States. Frontiers in Pharmacology. 10, Page 921.

20. US Food and Drug Administration, 2007. Overview of Device Regulations. www.fda.gov/cdrh/devadvice/overview.html.

21. Global Harmonization Task Force, 2008. http://www.ghtf.org.

22. US Food and Drug Administration, 2007. Standards Program, http://www.fda.gov/cdrh/stdsprog.html.

23. Läkemedelsverket, 2006. Introduktion till regelverket, http://www.lakemedelsverket.se.

24. Opderbeck DW et al, 2019. Artificial Intelligence in Pharmaceuticals, Biologics, and Medical Devices: Present and Future Regulatory Models. Fordham Law Review. 88(2), Pages 553–558.

25. TGA, 2005. Regulation of Therapeutic Goods in Australia. http://www.tga.gov.au/docs/html/tga/tgainfo.htm.

26. TGA, 2003. Australian Medical Devices Guidelines: An Overview of the New Medical Devices Regulatory System, Guidance Document Number 1, Version 1.6.

27. International Trade Administration, 2006. Medical Device Regulatory Requirements for Argentina. https://www.trade.gov/country-commercial-guides/argentina-medical-technology.

28. Iglesias Diez AM, et al, 2007. Registro de Productos Medicos, Demostracion de Seguridad y Eficacia, ANMAT.

29. ANVISA, 2008. Análise do Risco de Produtos Implantáveis Aspectos Legais e Técnicos.

30. ANVISA, 2007. Análise e Decisão de Petições de Tecnologia de Equipamentos Médicos. www.anvisa.gov.br.

31. Francielli Melo, 2008. Estudos Nao Clinicos e Clinicos, ANVISA.

32. Canadian Medical Devices Industry, 2005. 3.0 Medical Device Regulatory Framework, created.

33. Canadian Medical Devices Industry, 2007. Chapter 2: Canadian Requirements – An Overview of the Quality System Requirements for the Sale of Medical Devices in Canada.

34. Central Drugs Standard Control Organization, 2008. Clarification on Guidelines for Import and Manufacture of Medical Devices. https://cdsco.gov.in/opencms/opencms/en/Medical-Device-Diagnostics/Medical-Device-Diagnostics.

35. Central Drugs Standard Control Organization, 2005. Drugs and Cosmetic Act 1940, amended to May 30, 2005, at https://cdsco.gov.in/opencms/opencms/en/Notifications/documents/.

36. Pacific Bridge Medical, 2007. The New Market Authorization Holder (MAH) System for Medical Devices in Japan, AdvaMed.

37. Murayama Y et al, 2006. The Regulatory System and Requirements in Japan for Medical Devices, The New Pharmaceutical Affairs Law, TUV SUD, Japan.

38. International Trade Administration, 2002. Medical Device Regulatory Requirements for Mexico. https://www.trade.gov/market-intelligence/mexico-medical-devices.

39. Swedish Trade Council Mexico, 2007. Medical Devices Registration. www.swedishtrade.se/mexico.

40. Global Harmonization Task Force, 2000. Medical Devices Regulation in Mexico.

41. Pharmabitz, 2004. Import Registration and Taxation for Medical Equipment in Russia. http://www.pharmabiz.com/Services/ExportImport/Countries/Russia.aspx.

42. International Trade Administration, 2006. Medical Device Regulatory Requirements for Russia, Registering Medical Equipment in Russia.

43. International Trade Administration, 2008. Medical Device Regulatory Requirements for Korea.

44. Pharmacia, 2008. U.S. Device Makers Seek Exemptions from Korean DeviceTesting Rules. http://fdcalerts.typepad.com/asia/2007/07/us-device-maker.html.
45. Department of Health Taiwan (DOH), 2008. About DOH. https://www.ncbi.nlm.nih.gov/pmc/articles/PMC3960712/.
46. International Trade Administration, 2008. Medical Device Regulatory Requirements for Taiwan. https://www.trade.gov/country-commercial-guides/taiwan-medical-devices.
47. Pacific Bridge Medical, 2004. Taiwan Medical Device Regulation Update. www.pacificmedical.com.
48. Directive 2001/83/EC, Annex I, Part IV, as amended in Directive 107 2009/120/EC.
49. FDA. 1998. Guidance for Industry: Guidance for Human Somatic Cell Therapy and Gene Therapy.
50. Act on Pharmaceutical and Medical Devices, Chapter 1 Article 2–9*.
51. FDA. 2015. What Are "Biologics" Questions and Answers. https://www.fda.gov/aboutfda/centersoffices/officeofmedicalproductsandtobacco/cber/ucm133077.htm.
52. Information on CBER Restructuring. US Food and Drug Administration. 2016. https://www.fda.gov/AboutFDA/CentersOffices/.
53. Witten C, 2016. Office of Cellular, Tissue, and Gene Therapies Overview. PPMD Gene TherapyForum. https://www.parentprojectmd.org/research/current-research/our-strategy-impact/ppmds-gene-therapy-initiative.
54. BioPharm Dive, 2020. FDA, Expecting a Gene Therapy Boom, Firms Up Policies. https://www.biopharmadive.com/news/fda-gene-therapy-guidance-sameness-durability/571225/ (retrieved January 29, 2020).
55. Smith JA, Bravery CA, Hollander G et al, 2015. Regenerative Medicine Regulations: Cell Therapy, Gene Therapy and Tissue Engineering. Regulatory Affairs Professional Society.
56. Azuma K et al, 2015. Regulatory Landscape of Regenerative Medicine in Japan. Current Stem Cell Reports. 1(2), Pages 118–128.
57. Ecorys Nederland, 2016. Study on the Regulation of Advanced Therapies in Selected Jurisdictions. https://policycommons.net/artifacts/265359/study-on-the-regulation-of-advanced-therapies-in-selected-jurisdictions/1059412/.
58. Halioua-Haubolda, h C-L, Peyerb JG, Smithc JA et al, 2017. Regulatory Considerations for Gene Therapy Products in the US, EU, and Japan. Yale Journal of Biology and Medicine. 90, Pages 683–693.
59. Hayakawa akao, Harris I, Joung J et al, 2016. Report of the International Regulatory Forum on Human Cell Therapy and Gene Therapy Products. Biologicals. 44(5), Pages 467–479.

9 Artificial Intelligence in Community Pharmacy

Sriram Nagarajan
Holy Mary Institute of Technology and Science, Hyderabad, India

Kiruba Mohandoss
Sri Ramachandra Institute of Higher Education and
Research, Chennai, Tamil Nadu, India

Yoga Senbagapandian Rajamani
University of Maryland, College Park, Maryland, USA

Hanan Fahad Alharbi
Princess Nourah bint Abdul Rahman University,
Riyadh, Saudi Arabia

Muralikrishnan Dhanasekaran
Auburn University, Auburn, Alabama, USA

9.1 INTRODUCTION

Artificial Intelligence (AI) and machine learning–based technologies have the potential to transform healthcare because they offer new and important insights derived from the vast amount of data generated during the delivery of healthcare every day (1). Technology has begun playing a larger role in community pharmacy and the power of AI is poised to transform the world of pharmacy as we know it (2). The capacity of AI to learn from real-world feedback and improve its performance makes this technology uniquely suited as Software as a Medical Device (SaMD) and is responsible for it being a rapidly expanding area of research and development. Clinical pharmacy practice may undergo major change due to the implementation of this technology (3). The challenges facing clinical pharmacists include discovering how to apply AI technologies in ways that reveal new patterns in health data that make a real difference to clinical practice (4).

AI is expected to assist healthcare professionals in enhancing patient's experiences and health outcomes, augmenting the health of the population, reducing costs, and improving the interventions that pharmacists and other providers instigate with patients (5). Thus, the use of AI could help clinical staff to provide more informed medication-use decisions and improve outcomes. However, healthcare professionals must ensure that there is evidence indicating that any new SaMD to be implemented and all AI implementations are safe and effective before they are put into use in practice (6). Whether or not a Health Technology Assessment (HTA) process is performed, pharmacists clearly play a critical role in helping to generate the evidence that is needed to inform decisions concerning how and when to implement AI on a widespread basis in routine clinical pharmacy practice (7). Not all software used in the healthcare settings is considered to be a medical device. However, depending on its functionality and intended purpose, software may fall within the European Union (EU) definition of a "medical device" (8, 9)

DOI: 10.1201/9781003343981-9

The EU and the United States both have their own criteria for identifying healthcare and medical devices, although both definitions are the result of a purpose-based approach. AI software properly classified as a medical device must comply with the rules that aim to ensure its safety and level of performance. Given the capacity of AI to capture various forms of personal data, cybersecurity will also become very important to ensure the sustainability of this technology, including periodic reviews of the internal processes, to make sure it fulfills the requirements for the protection of privacy. In the EU, for example, the processing of personal data is governed by the General Data Protection Regulation (GDPR). Meanwhile, in the United States, regulatory issues may arise for AI developers based on the intended use of the product. Once a product is classified as a medical device, its class will define the regulatory requirements applicable for FDA clearance or approval. Regardless of the classification of the product (10), however, AI developers will need to assess whether the HIPAA (Health Insurance Portability and Accountability Act) rules apply, as well as any design controls and postmanufacture auditing that also apply in terms of cybersecurity (11). The traditional paradigm of medical device regulation was not designed for adaptive AI technologies, which have the potential to adapt and optimize device performance in real time to continuously improve healthcare for patients (12).

9.2 APPLICATIONS

The frontline of pharmacy is probably yet to feel the full force of the impact AI is making on the wider pharmaceutical industry. While facial recognition and speech pattern monitors can be used to detect rare diseases, it isn't like these systems are in operation in community pharmacies. Something that is more accessible is compliance technology (13). Though perhaps not in the guise that it's needed quite yet. Another accessible option for pharmacies is AI Sentiment Analysers, which are in a trial phases of a rollout for things like phone calls (14). "Here's the promise of technology: 62% of decision makers say AI will positively disrupt the pharmacy landscape within 1 to 2 years," Clark said. "Ninety-one percent of decision makers believe that technology can improve their pharmacies' profitability without question" (15). In terms of pricing, AI is able to compare the prices of major chain pharmacies with local competitors and identify the granular details of pricing. This information could assist pharmacies in establishing the most competitive prices, positioning themselves at the most competitive spot in the market (16). Because of the power and predictability of AI, many healthcare environments are starting to invest in similar technologies. It's especially important for community pharmacies to embrace AI and other technological advances to be able to compete with major chain pharmacies. As data and AI take (17) on a more pivotal role in healthcare, pharmacists will be equipped and empowered with the insights needed to assist in areas like medication reconciliation and adherence programs (18). These tools will help reshape the role of the retail pharmacist by expanding capabilities and responsibilities and generating awareness of what's already possible (19). When patients take more control over their health and become consumers in the healthcare market, customer demand for improved healthcare access and convenience are nonnegotiable (20). To this end, pharmacies must have the tools and technologies (21) required to meet customers where they are at the right time. Understanding data and how it can help them stay ahead of continually evolving trends and expectations will also be pivotal in the patient–pharmacist relationship (22).

REFERENCES

1. Mak KK, *Pichika* MR. Artificial intelligence in drug development: Present status and future prospects. Drug Discov Today. 2019;24(3):773–80.
2. Hassanzadeh P, Atyabi F, Dinarvand R. The significance of artificial intelligence in drug delivery system design. Adv Drug Deliv Rev. 2019;151:169–90.
3. Russel S, Dewey D, Tegmark M. Research priorities for robust and beneficial artificial intelligence. AI Mag. 2015;36(4):105–14.

4. Duch W, Setiono R, Zurada JM. Computational intelligence methods for rule-based data understanding. Proc IEEE. 2004;92(5):771–805.

5. Dasta JF. Application of artificial intelligence to pharmacy and medicine. Hosp Pharm. 1992;27(4):319–22.

6. Jiang F, Jiang Y, Zhi H. Artificial intelligence in healthcare: Past, present and future. Stroke Vasc Neurol. 2017;2(4):230–43.

7. Gobburu JV, Chen EP. Artificial neural networks as a novel approach to integrated pharmacokinetic–pharmacodynamic analysis. J Pharm Sci. 1996;85(5):505–10.

8. Sakiyama Y. The use of machine learning and nonlinear statistical tools for ADME prediction. Expert Opin Drug Metab Toxicol. 2009;5(2):149–69.

9. Agatonovic-Kustrin S, Beresford R. Basic concepts of artificial neural network (ANN) modeling and its application in pharmaceutical research. J Pharm Biomed Anal. 2000;22(5):717–27.

10. Mayr A, Klambauer G, Unterthiner T, Hochreither S. Deep Tox: Toxicity prediction using Deep Learning. Front Environ Sci. 2016;3:80.

11. Bishop CM. Model-based machine learning. Philos Trans A Math Phys Eng Sci. 2013;371(1984):20120222.

12. Merk D, Friedrich L, Grisoni F, Schneider G. *De novo* design of bioactive small molecules by artificial intelligence. Mol Inform. 2018;37(1–2):1–4.

13. Manikiran SS, Prasanthi NL. Artificial Intelligence: Milestones and Role in Pharma and Healthcare Sector. Pharma Times. 2019;51(1):10–1.

14. Cherkasov A, Hilpert K, Jenssen H, Fjell CD, Waldbrook M, Mullaly SC, et al. Use of artificial intelligence in the design of small peptide antibiotics effective against a broad spectrum of highly antibiotic resistant superbugs. ACS Chem Biol. 2009;4(1):65–74.

15. Achanta AS, Kowalski JG, Rhodes CT. Artificial neural networks: Implications for pharmaceutical sciences. Drug Dev Ind Pharm. 1995;21(1):119–55.

16. Sakiyama Y. The use of machine learning and nonlinear statistical tools for ADME prediction. Expert Opin Drug Metab Toxicol. 2009;5(2):149–69.

17. Heikamp K, Bajorath J. Support vector machines for drug discovery. Expert Opin Drug Discov. 2014;9(1):93–104.

18. Sutariya V, Groshev A, Sadana P, Bhatia D, Pathak Y. Artificial neural network in drug delivery and pharmaceutical research. Open Bioinf J. 2013;7(1):49–62.

19. Gutiérrez PA, Hervás-Martínez C. Hybrid artificial neural networks: Models, algorithms and data. Advances in Computational Intelligence, Lecture Notes in Computer Science, Springer, Berlin, Heidelberg. 2011;6692.

20. Fleming N. How artificial intelligence is changing drug discovery. Nature. 2018;557(7706):S55–7.

21. Sun Y, Peng Y, Chen Y, Shukla AJ. Application of artificial neural networks in the design of controlled release drug delivery systems. Adv Drug Deliv Rev. 2003;55(9):1201–15.

22. Manda A, Walker RB, Khamanga SMM. An artificial neural network approach to predict the effects of formulation and process variables on prednisone release from a multipartite system. Pharmaceutics. 2019;11(3):109.

10 An Ethical Outlook on the Applications of Artificial Intelligence in Medicine and Pharmacy

Akila Ramanathan
College of Pharmacy, Sri Ramakrishna Institute of
Paramedical Sciences, Coimbatore, Tamil Nadu, India

K. Reeta Vijaya Rani
Surya School of Pharmacy, Vikravandi, Tamil Nadu, India

Mullaicharam Bhupathyraaj
College of Pharmacy, National University of
Science and Technology, Muscat, Oman

Leena Chacko
Meso Scale Diagnostics LLC, Rockville, Maryland, USA

Hanan Fahad Alharbi
Princess Nourah bint Abdul Rahman University,
Riyadh, Saudi Arabia

10.1 INTRODUCTION

In the pharmaceutical sector, Artificial Intelligence (AI) is being investigated at three stages: pre-clinical to better understand disease biology and drug candidates; clinical to assist in selecting populations and planning intervention studies; and digital treatments and devices to enable continuous monitoring. Thus, the pharmaceutical value chain allows organizations to capture valuable data at every touchpoint, from biomedical research to drug production, via patient and healthcare community interaction. Using AI at scale enables the industry to capitalize on the value of this data to get critical insights (1). It is a game changer in the pharmaceutical sector because of its capacity to use these insights to inform and expedite decision-making (2).

AI could also assist healthcare practitioners in avoiding errors, allowing doctors to focus on giving therapy and addressing difficult problems. The potential benefits of these technologies, as well as the economic and commercial opportunities for AI in healthcare, suggest that AI will be employed more widely around the world. However, it is a two-edged sword. AI-enabled products, for example, have occasionally resulted in erroneous, even potentially hazardous, therapy recommendations (3). As algorithms are involved, all AI judgments, even the quickest, are systematic, in contrast to human decision-making. As a result, even if activities do not have legal consequences, they inevitably lead to accountability, not by the machine but by the people who constructed it and use it. We must be aware that its disadvantages may outweigh its advantages.

DOI: 10.1201/9781003343981-10

Many ethical issues endangering the health and human rights remain unresolved. Melanoma-identifying AI algorithms are now trained mostly on photos of white skin, rendering them ineffective at detecting melanoma in darker-skinned individuals (despite melanoma being more fatal in African populations) (4). If ethical challenges like these are not addressed quickly, we risk an "AI winter," in which public trust and the potential benefits of AI for healthcare are lost rapidly (5). To solve this dilemma, professionals must consider humanity and ethics. Software developers, healthcare professionals, politicians, and patients all have a role to play in tackling these diverse difficulties. Regulatory bodies may also need to modify their current supervision techniques to keep up with the rapid changes in this industry (6). Although regulatory legislation to address these challenges at the legal level (7) and governance frameworks to handle these issues at the organizational level (8) are steadily evolving, there are still a few practical guidelines that developers and users of AI for healthcare can follow. Creating and publishing an AI and ethics policy can help life sciences companies and healthcare providers build trust in the technologies they utilize (9).

10.2   AI

The term "artificial intelligence" generally refers to computer programs performing tasks that are commonly associated with intelligent beings. Algorithms are the foundation of AI; they are translated into computer code that contains instructions for rapid data analysis and transformation into conclusions, information, or other outputs.

Massive amounts of data, as well as the ability to rapidly analyze such data, power AI (10). According to the Organization for Economic Co-operation and Development (OECD) Council on AI recommendation (11), an AI system is a machine-based system that can, for a given set of human-defined objectives, make predictions, recommendations, or decisions influencing real or virtual environments. AI systems are built with varying degrees of autonomy in mind.

Machine learning (ML) applications such as pattern recognition, natural language processing (NLP), signal processing, and expert systems are examples of AI technology. ML, a subset of AI techniques, is based on the definition and analysis of data using statistical and mathematical modeling techniques. These patterns are then used to perform or guide specific tasks and make predictions.

ML is classified into three types based on how it learns from data: supervised learning, unsupervised learning, and reinforced learning. The data used to train the model is labeled (the outcome variable is known) in supervised learning, and the model infers a function from the data that can be used to predict outputs from different inputs. Unsupervised learning involves a machine identifying hidden patterns in data rather than labeling it. Reinforcement learning entails ML through trial and error to achieve a goal, for which the machine is "rewarded" or "punished" depending on whether its inferences help or hinder achievement of the goal (12).

Deep learning, also known as "deep structured learning," is a type of ML that employs multilayered models to extract features from data progressively. Deep learning can be both supervised and unsupervised. Deep learning models are typically fed large amounts of data. Many ML approaches are data-driven. They depend on large amounts of accurate data, referred to as "big data," to produce tangible results. "Big data"are complex data that are rapidly collected in such unprecedented quantities that terabytes (1 trillion units [bytes] of digital information), petabytes (1,000 terabytes), or even zettabytes (1 million petabytes) of storage space may be required as well as unconventional methods for their handling.

AI has the potential to improve healthcare delivery in areas such as disease prevention, diagnosis, and treatment (13). Usable data in healthcare has proliferated as a result of collection from various sources, including wearable technologies, genetic information generated by genome sequencing, electronic healthcare records, radiological images, and even hospital rooms.

10.3 APPLICATIONS OF AI IN HEALTHCARE

AI in healthcare has the potential to solve significant challenges in the field of healthcare such as therapeutics, diagnosis and screening, preventive treatments, clinical decision-making, public health surveillance, complex data analysis, and disease prediction. This list is likely to grow indefinitely.

10.3.1 THERAPEUTICS/RESEARCH AND DRUG DISCOVERY

AI and ML are being used in drug discovery and epitope identification for vaccine development, with the potential to speed up and reduce costs. Precision medicine, as the name implies, investigates the possibility of providing individualized treatments based on an individual's unique characteristics, such as age, gender, race, family history, and genomic variation. Large datasets, such as genomic, sociodemographic, and electronic medical records (EMR), can be used by ML algorithms to predict disease outcomes (14). AI technology can guide treatment plans by using genetic analysis and personalized drugs to target specific health conditions.

10.3.2 DRUG DEVELOPMENT

AI has the potential to improve the R&D process. AI can do everything from designing and identifying new molecules to target-based drug validation and discovery. According to an MIT study, only 13.8% of drugs pass clinical trials. To top it all off, a pharmaceutical company must pay anywhere from US$161 million to US$2 billion for a drug to go through the entire clinical trial process and receive FDA approval. These are the two main reasons why pharmaceutical companies are increasingly embracing AI—to improve the success rates of new drugs, to develop more affordable drugs and therapies, and, most importantly, to reduce operational costs.

10.3.3 DIAGNOSTICS AND SCREENING

The AI application gives a competitive advantage in disease detection and offers the necessary hope to alleviate the diagnosis and screening burden on the healthcare system. According to the National Academies of Sciences, Engineering, and Medicine report, "post-mortem studies have shown that around 10% of patient deaths can be attributed to diagnostic errors," and they also reported that diagnostic errors account for 6–17% of adverse events based on a review of medical records (15). As a result, AI-based technologies may help reduce human error in healthcare.

AI-based tools have the potential to improve known methods of disease screening and diagnosis, improve diagnostic accuracy, and guide evidence-based treatment algorithms, predict outcomes, and identify health system gaps, all with an impact on human health and well-being. Recently, AI technology has been used to predict genetic makeup based on body phenotypes.

10.3.4 CLINICAL CARE

Healthcare demand is increasing, and countries are facing a skilled labor shortage. AI advancements have created new opportunities to address this shortage. Telemedicine and self-care via interactive chatbots, as well as digital monitoring devices such as wearables, are two areas that have seen significant advancement in recent years. This also gives healthcare workers another option for remote monitoring and detecting early signs of disease (16).

To aid clinical decision-making, NLP is being used to analyze unstructured data such as physician–clinical notes (17). Google Deep Mind and IBM Watson Analytics have developed AI-powered tools for improving overall patient outcomes, such as mobile-based medical assistants, diagnostic tools, clinical decision-making tools, and prognostic prediction tools (18, 19). AI technology can aid in the self-monitoring of personal health-related parameters such as nutrition intake, physical activity, blood pressure, glucose, and lipids in order to identify high-risk individuals. AI-powered health coaching systems and smartphone apps that use neural networks and ML methods could help with medication adherence, motivation, reminders, and building a care network (20). Chatbots and robotic assistants can help patients manage noncommunicable diseases (NCDs) on their own.

10.3.5 EPIDEMIOLOGY AND DISEASE PREVENTION

Epidemiology is the bedrock of public health, guiding policy decisions and evidence-based practice. The science entails identifying disease factors and determinants, as well as trends, patterns, and disease prediction. Conventional data collection methods involve one or two sources, but AI methods have the potential to integrate data from multiple sources, including surveillance, administrative, hospital data, registries, and General practitioner (GP) clinics, to provide meaningful evidence. AI and ML tools enable the efficient and accurate handling of large and diverse datasets, as well as data-driven solutions for risk prediction and risk mitigation.

For example, during the early stages of the COVID-19 pandemic, many countries used AI-based methods for early detection and contact tracing to monitor the disease's spread (21). Deep learning algorithms trained to read COVID-19 CT scans performed better. AI solutions, such as medical image interpretation, societal, behavioral, and health data analysis, and medical record analysis, can provide a decision support system for both individuals and large-scale preventive intervention planning. It can aid in the reduction of risk factors and hazardous exposures in locations by utilizing Geographic Information System (GIS)-based sources and automation services.

10.3.6 BEHAVIORAL AND MENTAL HEALTHCARE

The medical AI model opens up new avenues for behavioral and mental health treatment. Medical AI has the potential to improve psychology and psychiatric procedures in a variety of ways, such as assisting patients in receiving a diagnosis, actively managing their symptoms between in-person consultations, predicting and preventing likely flare-ups, and more.

Individuals suffering from a variety of mental and behavioral disorders exhibit distinct symptoms that can be identified through verbal output (written or spoken), facial expressions, tone of voice, body language, and a variety of other factors. AI psychology and psychiatry models will help patients maintain active self-care in order to ensure better treatment of their illnesses and optimal mental and behavioral health.

Chatbots are one promising application of AI in mental health (22). While mental illnesses continue to carry a significant social stigma, and many people struggle to express their thoughts and feelings directly, mental health chatbots allow people who are hesitant to seek direct professional psychological and psychiatric help to take their first step toward self-care. In this approach, cognitive AI chatbots can provide initial help to people who aren't ready for professional or non-professional care, as well as supplement that helps in-between interactions with psychologists, psychiatrists, and peers.

10.3.7 REMOTE SURVEILLANCE

Remote surveillance is a game changer in the pharmaceutical and healthcare industries. Many pharmaceutical companies have already developed wearables that use AI algorithms to remotely monitor patients with life-threatening diseases.

Tencent Holdings, for example, has worked with Medopad to create an AI technology that can remotely monitor patients with Parkinson's disease, cutting the time required to perform a motor function assessment from 30 minutes to 3 minutes. It is possible to monitor the opening and closing motions of a patient's hands from a remote location by integrating this AI technology with smartphone apps.

When the smartphone camera detects hand movement, it records it in order to determine the severity of the symptoms (Parkinson's). The movement's frequency and amplitude will change.

The frequency and amplitude of the movement will determine the severity score of the patient's condition, allowing doctors to remotely change the drugs and drug doses (23).

10.3.8 MANUFACTURING

Pharma companies can use AI in the manufacturing process to increase productivity, enhance efficiency, and expedite the production of life-saving drugs. AI can be used to manage and improve all aspects of the manufacturing process:

Quality assurance
Maintenance that is predicted
Waste minimization
Design enhancement
Automation of processes

AI can replace time-consuming traditional manufacturing techniques, allowing pharmaceutical companies to launch drugs much faster and at a lower cost. AI would not only significantly increase their return on investment (ROI) by limiting human intervention in the manufacturing process, but it would also eliminate any possibility of human error (24).

10.3.9 MARKETING

Given the pharmaceutical industry's emphasis on sales, AI can be a useful tool in pharma marketing. Pharma companies can use AI to experiment with and develop novel marketing strategies that promise high revenue and brand awareness.

AI can assist in mapping the customer journey, allowing businesses to see which marketing technique drove visitors to their site (lead conversion) and ultimately pushed the converted visitors to buy from them. Pharma companies can then focus more on marketing strategies that result in the most conversions and revenue growth (25). Integration and adoption of AI necessitate industry expertise and skills, which are currently in short supply. However, the process of AI adoption in the pharmaceutical industry can be simplified by collaborating and partnering with academic institutions that specialize in AI R&D to help pharma companies with AI adoption, collaborating with firms that specialize in AI-driven medicine discovery to benefit from expert advice, cutting-edge tools, and industry experience; and training R&D and manufacturing teams on how to use and implement AI tools and techniques for maximum productivity.

10.3.10 DRUG ADHERENCE AND DOSAGE

The use of AI in pharmacy is expanding at an unprecedented rate. AI in pharma is now being used to determine the appropriate amounts of drug intake to ensure the safety of drug consumers. It not only helps to monitor patients during clinical trials, but it also suggests the appropriate dosage at regular intervals (26).

AI in pharmaceuticals has resulted in faster process automation and is one of the key factors driving the increasing need for accuracy in this industry. The opportunities for AI in pharma are immeasurable, ensuring both efficiency and compliance. Furthermore, AI in pharmaceuticals has unlocked several potential AI jobs for people, which come with lucrative salaries and benefits.

10.3.11 SYSTEM OF HEALTH MANAGEMENT USING AI

AI has the potential to improve and optimize operational functions in a healthcare setting or organization. Scheduling, admission, EMR, accounting, billing, and claim settlement all are aspects of healthcare management that involve repetitive tasks and a high level of scrutiny. Productivity could be increased, operational and clinical workflows could be improved, and healthcare practices' operating costs could be reduced by leveraging AI-powered tools and automated processes. Robotic process automation (RPA) can handle advanced financial accounting, medical billing, and claims processing. NLP has the potential to automate clinical documentation, reducing turnaround time. Inpatient and outpatient scheduling, interdepartmental coordination, and patient alerts all can be improved with AI healthcare administration tools. Thus, AI technology could be beneficial in both patient care and back-office operations, enhancing efficiency in the health sector (27).

10.3.11.1   Medical AI Software for Clinic Management Systems

Medical AI has the potential to increase the autonomy, efficiency, and functionality of clinic management systems (CMS). Custom machine language (ML) is quickly being used to complete tasks previously completed by employees. To match the redirected processes, task-scheduling and appointment-making can be automatically modified to meet changing conditions, with assignments and time tables adjusted on the fly and notifications sent to relevant physicians and other staff. The CMS model also makes it simpler for clinicians to store EMR and for patients to access it, allowing for faster access and preservation of health and treatment histories for faster decision-making based on a more comprehensive understanding of the patient's unique health profile.

10.3.11.2   Medical AI Software for Hospital Management Systems (HMS)

Medical AI, like CMS, can improve HMS. Hospitals face unique challenges that may offer greater potential for improved performance through the use of AI technology. Medical AI may help with inpatient and outpatient scheduling, where decisions on patient rotation can be made based on a variety of parameters, including prognosis, prior health history, treatment response, available personnel, and more. AI HMS systems can facilitate interdepartmental communication and coordination to ensure the best possible use of resources and time for all patients, with alerts issued when specific areas are stressed to provide direction and a head start on finding alternative solutions as needed.

10.4   LAWS AND POLICIES, CONCERNING AI FOR HEALTH

10.4.1   HUMAN RIGHTS AND AI

The tension between AI and human rights is becoming more apparent as technology becomes more central to our daily lives and the functioning of society. While AI is viewed as a benefit to modern society, the lack of stringent data protection policies provides tech companies with a society primed for digital exploitation. With little oversight or accountability, these corporations freely intrude into citizens' lives and increasingly violate human rights. AI has proven to be a threat to equal protection, economic rights, and fundamental freedoms, from fostering discrimination to invasive surveillance practices. In order to revert these trends, appropriate legal standards must be implemented in our digitally transforming societies. Transparency in AI decision-making processes, improved accountability for tech titans, and the ability for civil society to challenge the implementation of new technologies in society are all desperately needed. AI literacy should also be promoted through investments in public awareness and education initiatives that help communities learn not only about the functions of AI, but also about its impact on our daily lives. Unless appropriate measures are implemented to protect the interests of human society, the future of human rights in this technological era remains uncertain (28).

The Ad-hoc Committee on Artificial Intelligence was established by the Council of Europe in 2019–2020 to conduct broad multistake holder consultations to determine the feasibility and potential elements of a legal framework for the design and application of AI in accordance with the Council of Europe's standards on human rights, democracy, and the rule of law. Furthermore, in 2019, the Council of Europe issued Guidelines on Artificial Intelligence and Data Protection (29), which are based on the protection of human dignity as well as the safeguarding of human rights and fundamental freedoms. Furthermore, the European Commission for Efficiency in Justice's ethical charter includes five principles relevant to the use of AI for health (30).

10.4.2 DATA PROTECTION LEGISLATION AND POLICIES

Data protection laws are "rights-based approaches" that establish standards for regulating data processing that protect individuals' rights while also establishing obligations for data controllers and processors. Data protection laws are also increasingly acknowledging that people have the right not to be subjected to decisions that are solely guided by automated processes. Over 100 countries have passed data protection legislation. The General Data Protection Regulation (GDPR) of the European Union is one well-known set of data protection laws; in the United States, the Health Insurance Portability and Accountability Act, enacted in 1996, governs health data privacy and security (31, 32). Some standards and guidelines are designed specifically to manage the use of personal data for AI. For example, the Ibero-American Data Protection Network, comprised of 22 data protection authorities from Portugal and Spain, as well as Mexico and other countries in Central and South America and the Caribbean, has issued general recommendations for the processing of personal data in AI (33) as well as specific guidelines for compliance with the principles and rights which govern the cybersecurity in data science (34).

10.4.3 BIOETHICS LEGISLATION AND POLICIES

Several bioethics laws and policies have been revised in recent years to reflect the growing use of AI in science, healthcare, and medicine. It includes human supervision standards, also known as "human warranties," which require patient and clinician evaluation at key points in the development and deployment of AI. It also supports the development of a secure national platform for health data collection and processing, as well as free and informed consent for data use (35).

10.5 ETHICAL GUIDELINES FOR AI TECHNOLOGY IN HEALTHCARE

Ethical principles for the application of AI to health and other domains are intended to guide developers, users, and regulators in improving and monitoring the design and use of such technologies. Human dignity and inherent worth are the pillars upon which all other ethical principles are built. These principles are patient-centric and are expected to guide all the stakeholders in the development and deployment of responsible and reliable AI for health. These principles are as follows.

10.5.1 AUTONOMY

When AI technologies are used in healthcare, there is a risk that the system will function autonomously, undermining human autonomy. The application of AI technology to healthcare may place decision-making power in the hands of machines. The AI-based healthcare system and medical decision-making should be completely controlled by humans. Under no circumstances should AI technology interfere with patient autonomy.

The "Human in the Loop" (HITL) (36) model of AI technologies allows humans to supervise the system's operation and performance. Clinical decisions made by AI technology and physicians may differ, causing the user/patient to be unsure whether to trust the clinician or the AI technology. The patient should be given both options in such cases. Before introducing any AI technology into healthcare, patients must be fully informed about the use of AI technologies, their benefits, and associated physical, psychological, and social risks. Patients must have complete discretion in deciding whether or not to use AI technologies. Human rights must be effectively and transparently monitored.

The patient or participant has the right to refuse consent. The government, sponsors, researchers, healthcare professionals, or other stakeholders should not be forced to use such AI technologies. Overreliance on AI systems for diagnosis and treatment may have a negative impact on the patient–clinician relationship and the patient's autonomy. As a result, developers, institutions, hospitals, health systems, and other stakeholders should develop policies and guidelines to increase participants' autonomy.

10.5.2 Encourage Human Well-Being, Safety, and the Public Benefit

People should not be harmed by AI. Before deployment, they must meet regulatory requirements for safety, accuracy, and efficacy, and measures to ensure quality control and improvement must be in place. As a result, funders, developers, and users have a continuing obligation to measure and monitor the performance of AI algorithms in order to ensure that AI technologies work as intended and to determine whether they have any negative impact on individual patients or groups (37).

To avoid harm, AI technologies must not cause mental or physical harm. AI technologies that provide a diagnosis or warning that an individual cannot address due to a lack of appropriate, accessible, or affordable healthcare should be carefully managed and balanced against any "duty to warn" that may arise from incidental and other findings, and appropriate safeguards should be in place to protect individuals from stigma or discrimination based on their health status.

10.5.3 Maintain Transparency, Explainability, and Intelligibility

AI should be understandable or intelligible to developers, users, and regulators.

Improving the transparency and explainability of AI technology are two broad approaches to ensuring intelligibility. Transparency necessitates the publication or documentation of sufficient information prior to the design and deployment of an AI technology. Such information should enable meaningful public consultation and debate about how AI technology should be designed and used. After an AI technology is approved for use, such information should be published and documented on a regular and timely basis.

Transparency will boost system quality while also safeguarding patient and public health and safety. System evaluators, for example, require transparency to identify errors, and government regulators rely on transparency to conduct proper, effective oversight. An AI technology must be auditable, including when something goes wrong. Transparency should include accurate information about the technology's assumptions and limitations, operating protocols, data properties (including data collection, processing, and labeling methods), and algorithmic model development.

All algorithms should be rigorously tested in the environments where the technology will be used to ensure that it meets safety and efficacy standards. Such tests and evaluations should be subject to rigorous, independent oversight to ensure that they are carried out safely and effectively. Healthcare institutions, health systems, and public health agencies should regularly publish information on how decisions for AI technology adoption were made, how the technology will be evaluated on a regular basis, its uses, known limitations, and the role of decision-making, which can facilitate external auditing and oversight (37).

10.5.4 Data Security

Access to patient medical data is frequently critical to the application of AI in healthcare delivery. As the use of AI products to share medical information between patients, physicians, and the care team grows, preserving an individual's confidentiality and privacy becomes ever more vital. Entities utilizing or selling AI-based healthcare goods must consider federal and state rules and regulations that regulate the preservation and use of patient information, as well as other typical practical difficulties that face AI-based healthcare products. The gathering and use of patient health information in AI products may inadvertently violate the Health Insurance Portability and Accountability Act (HIPAA) and many state privacy and security regulations and laws. It is critical for AI healthcare enterprises and institutions employing AI healthcare products to determine if HIPAA or other state rules apply to the data. Before entrusting any third party with patient data, including protected health information (PHI), it is critical to perform adequate vendor due diligence. Two critical factors for due diligence are how the data is gathered (e.g., straight from medical records) and where it is finally stored. Failure to do proper due diligence in either situation might have legal and monetary implications. Appropriate security protections must be implemented to protect privacy while also increasing confidence in the technology. Information systems and data should be examined and monitored on a regular basis to detect any data breaches. It is critical to establish who will have access to the data and algorithms and to implement stringent restrictions suitable for the amount of access. Personnel and vendors must be informed of their access constraints, data usage limitations, and data security requirements. Such advancements may be transient unless patients and physicians have confidence that these AI-based companies are considering and protecting patient data privacy (38).

10.5.5 ENCOURAGE RESPONSIBILITY AND ACCOUNTABILITY

Although AI technologies can perform specific tasks, it is the responsibility of stakeholders to ensure that they can perform those tasks and that they are used appropriately.

The application of "human warranty," which entails evaluation by patients and clinicians in the development and deployment of AI technologies, can ensure accountability. Regulatory principles are applied upstream and downstream of the algorithm in human warranty by establishing points of human supervision. Discussions among professionals, patients, and designers identify critical points of supervision. The goal is to keep the algorithm on a ML development path that is medically effective, interrogable, ethical, and accountable; it entails an active collaboration with patients and the public, such as meaningful public consultation and debate (39).

When something goes wrong with the application of AI technology, someone should be held accountable. Appropriate mechanisms should be implemented to ensure accountability and redress for individuals and groups harmed by algorithmically informed decisions. This should include prompt, effective remedies and redress from governments and companies that use AI in healthcare. Redress should include compensation, rehabilitation, restitution, and, if necessary, sanctions, as well as a guarantee of non-repetition. The use of AI technologies in medicine necessitates the assignment of responsibility. These are complex systems with multiple agents sharing responsibility. When AI-powered medical decisions cause harm to individuals, responsibility and accountability processes must clearly define the relative roles of manufacturers and clinical users in the harm.

10.5.6 ENSURE INCLUSIVENESS AND EQUITY

To be inclusive, AI for health must be intended to support the broadest possible equitable use and access, regardless of age, gender, poverty, race, ethnicity, sexual orientation, ability, or other human rights–protected traits. To prevent hurdles from using, AI developers and vendors should consider the diversity of languages, abilities, and modes of communication around the world. Industry and governments should work together to prevent the "digital divide" from widening inside and across countries, and to ensure equitable access to breakthrough AI technologies.

AI technologies should not be prejudiced, and they threaten inclusion and fairness since they constitute a divergence from equal treatment, which is frequently arbitrary. AI developers must ensure that AI data, particularly training data, is free of sample bias and accurate, complete, and diverse. If a particular racial or ethnic minority (or other groups) is neglected in a dataset, it may be required to oversample that group relative to its population size to ensure that an AI system obtains the same quality of findings in that population as in better-represented groups. AI technology should help to reduce the inherent power imbalances that exist between physicians and patients, or between firms that develop and implement AI technologies and those that use or rely on them (37).

10.5.7 Promoting Responsive and Sustainable AI

Designers, developers, and users should evaluate AI applications on a regular and transparent basis to determine if AI reacts successfully and correctly to expectations and needs. When an AI technology is unsuccessful or causes dissatisfaction, the obligation to respond necessitates an institutional mechanism to address the issue, which may include discontinuing the usage of the technology. Responsiveness also necessitates that AI technology is consistent with broader initiatives to promote healthcare systems, environmental sustainability, and workplace sustainability. AI technologies should be implemented only if they can be properly integrated and sustained in the healthcare system. AI systems should also be built to have a minimal environmental impact and maximize energy efficiency. Both governments and companies should prepare for anticipated workplace disruptions, such as training for healthcare employees to adjust to the use of AI systems and probable job losses owing to the usage of automated systems (40).

10.6 GUIDELINES FOR STAKEHOLDERS

The creation, validation/testing, and implementation of AI-based healthcare solutions is a multistep process involving partners from several fields of expertise. To ensure that AI-based solutions are technically and morally sound, each of these phases must adhere to accepted principles. With justice and fairness, the policy is justifiable and relevant to a vast number of people. All stakeholders must follow these guiding principles in order to make technology more helpful and acceptable to users and beneficiaries.

10.6.1 Guidelines for the Development Phase

The whole idea of design and product development process should be based on feedback from all stakeholders, including health professionals, to ensure that the finished product functions as intended. The following factors should be considered when gathering data:

- The aim and ultimate goal of data collecting by AI technology developers should be explained to hospitals, institutes, technicians, and AI technology developers.
- If applicable, pre-consent should be obtained. The data determines informed consent and reconsent.
- The information should not be used to cause harm or discriminate against anybody.
- To preserve people's privacy, AI technology developers should employ measures such as data encryption and data anonymization.
- There should be a feedback mechanism for which the user/physician can communicate concerns and suggestions with the developers.
- The organizations or the researchers must be truthful to the participants as to how their health data will be used.
- There should be appropriate provisions for disciplinary (legal or financial) actions in case the providers fail to comply with these regulations. The relevant stakeholders should be made liable to pay compensation to the users in case of any harm or injury arising from the use of AI technologies.
- There should be a feedback system in place to allow users and doctors to voice their problems and ideas to the developers.

- Organizations or researchers must be open and honest with participants about how their health data will be utilized.
- If providers fail to comply with these requirements, there should be suitable measures for disciplinary (legal or financial) actions. The key stakeholders should be held responsible for compensating consumers in the event of harm or injury caused by the usage of AI technology.

10.6.2 GUIDELINES FOR VALIDATION

- To validate the performance of an AI technology, datasets other than the training dataset should be employed. Algorithm auditing has also been offered as a feasible tool for the rigorous validation of AI (41).
- Clinical validation evaluates the clinical accuracy of AI-based solutions for a given clinical condition. The approach must also determine if the suggested AI-based solution is appropriate for the purpose.
- Robust validation of an AI system necessitates a multidimensional, multi-sectoral team comprising clinical, data science, statistical, engineering, public health, and epidemiological expertise.
- The validation should be impartial and based on both scientific research and medical concepts.
- In addition to accuracy, the validation process should incorporate usability and user experience findings. It should also contain a risk assessment for healthcare professionals and recipients.
- AI technology output, as well as the process for interpreting the outputs, should be comprehensible.

AI technology validation is continually improving, with newer and more complicated statistical indicators emerging to evaluate the effectiveness of AI for health. Validation studies should use appropriate and up-to-date analytics such as AUCROC, sensitivity, specificity, F1 score, and Matthew's correlation coefficient (42, 43).

10.6.3 GUIDELINES FOR CLINICAL AND OTHER HEALTH-RELATED DEPLOYMENT

AI technology deployment should be approached with extreme caution since poorly planned healthcare AI deployment can have a major detrimental impact on patients and healthcare practitioners.

- Whether the data is utilized for research, patient treatment, or public health decision-making, interested health professionals should be informed of how the information gathered from participants will be used. Healthcare practitioners should be properly trained in the appropriate and safe application of AI technologies.
- To achieve optimal performance, the AI technologies should be assessed in a range of challenging settings during the first deployment phase.
- Prior to implementation, health practitioners should have a good understanding of the AI technology's functional foundation. This should also contain a SWOT analysis of the proposed solution's strengths, weaknesses, opportunities, and threats (SWOT).
- Before clinical deployment, the AI technology user should be trained on the many types of bias that may emerge once the tool is employed (44).
- The terms of service should include links informing end users that the technology is AI and that the findings, diagnoses, and interpretations are not done by humans. In this regard, a disclaimer should be clearly communicated to the health professional and the individual for whom the technology has been used.

10.7 ETHICAL COMMITTEE REVIEW OF MEDICAL AI

When gaining research authorization, all materials are reviewed by a study site's local (independent) ethics committee (IEC). Prospective studies are frequently designed, and ethical committees play an essential part in their execution. The IEC may refuse to authorize the research based on the study results. For example, if interim safety assessments reveal patient safety hazards, the national ethics council can halt the study at the location.

AI studies are featured in a distinct dossier block. A retrospective design utilizing existing datasets is appropriate for most AI-based investigations.

Ethics difficulties are not precisely controlled in AI-based investigations; however, ethical committees are frequently informed of planned retrospective studies. Although this strategy does not violate any applicable rules or regulations, it does include some hazards that should be minimized. When working with a patient database, it is vital to foresee the dangers of data breaches and personal patient identification, as well as gather adequate high-quality data for subsequent research. Each IEC expert has more than five years of research experience and has been trained in good clinical practice. IEC evaluates study designs, including retrospective designs, while paying close attention to data security issues, data use and potential risks, the data collection process, and data storage and processing.

Because of the rapid scientific advancement and broad options for AI system application, deeper methodologies for ethical analysis and risk assessment should be considered when designing such investigations. The ethical committee is responsible for peer evaluation of the scientific work of AI-based medical investigations in addition to reviewing studies. The IEC produces a review and a set of suggestions for each work, assisting researchers in avoiding mistakes before submitting papers for publication. As a result, the IEC serves as an extra scientific supervisory body, assisting authors in improving the quality of their papers for top-tier journals.

The world's premier scientific journal will not examine articles unless the ethics committee has approved them. Protecting patient rights, data security, confidentiality, and the legality of materials submitted for publication are all top priorities in today's scientific world. Not only are researchers accountable for data veracity, but also the ethical boards that authorize experiments and evaluate articles.

IEC specialists thoroughly examine documentation, including at least a research strategy, patient-informed consent, and the lead investigator's professional vitae. To prevent errors in the study's design and interpretation, in addition to a thorough examination of the papers presented, the ethics committee offers suggestions for strengthening articles for subsequent publication in prominent journals. The IEC provides ethical support for medicinal products and medical device scientific and clinical trials, as well as assistance in the conduct of high-quality clinical and scientific investigations (45).

10.8 INFORMED CONSENT PROCESS

In healthcare, the basic concept of informed consent seems fairly straightforward. A patient is informed about a proposed test, treatment, or procedure; its benefits and risks; and any alternative options. With this knowledge, the patient decides to either consent or not consent to the recommended plan. In reality, though, informed consent is a more complex process that involves nondelegable duties and varies in scope based on the type of test, treatment, or procedure involved.

As AI advances into healthcare with applications such as ML, deep learning, neural networks, and NLP, new ethical and practical issues related to informed consent are emerging. AI's rapid momentum has, in many cases, eclipsed the ability of regulators, leaders, and experts to implement laws, standards, guidelines, and best practices that address some of these issues. "When an AI device is used, the presentation of information can be complicated by possible patient and physician fears, overconfidence, or confusion." "For an informed consent process to proceed appropriately, it requires physicians to be sufficiently knowledgeable to explain to patients how an AI device works" (46). Although acquiring extensive knowledge of AI coding, programming, and functioning is likely unrealistic for most healthcare providers, those who plan to use these technologies in practice should be able to

- provide patients with a general explanation of how the AI program or system works;
- explain the healthcare provider's experience using the AI program or system;
- describe to patients the risks versus potential benefits of the AI technology (e.g., compared to human accuracy);
- discuss with patients the human versus machine roles and responsibilities in diagnosis, treatment, and procedures;
- describe any safeguards that have been put in place, such as cross-checking results between clinicians and AI programs; and
- explain issues related to confidentiality of patient's information and any data privacy risks.

Taking the time to provide patients with these additional details during the informed consent process and to answer any questions can help ensure that they have the appropriate information to make informed decisions about their treatment. Following the informed consent process, providers should document these discussions in patients' health records and include copies of any related consent forms.

10.9   REGULATORY CONSIDERATIONS

The FDA is in charge of assuring the safety and efficacy of numerous AI-driven medical devices. The government generally controls software based on its intended purpose and potential danger to patients if it is wrong. The FDA classifies software as a medical device (SaMD) designed to treat, diagnose, cure, alleviate, or prevent illness or other disorders (47). The majority of medical devices that use AI or ML are classified as SaMD (48). Examples of SaMD include software that analyzes MRI images to detect and diagnose strokes and computer-aided detection (CAD) software that analyzes pictures to aid in the diagnosis of breast cancer (49). Some consumer-facing goods, such as certain smartphone applications, may also be categorized as SaMD.

10.9.1   FDA Recommendations

The Food and Drug Administration (FDA), Health Canada, and the Medicines and Healthcare Products Regulatory Agency (MHRA) of the United Kingdom have developed ten guiding principles that can help influence the development of good ML practices (GMLP). These guiding principles will aid in the promotion of medical devices that incorporate AI and ML that are safe, effective, and of high quality:

- *Throughout the Product Life Cycle, Multidisciplinary Expertise Is Leveraged:* An in-depth understanding of a model's intended integration into clinical workflow, as well as the expected benefits and associated patient risks, may assist guarantee that ML-enabled medical devices are safe and effective, and fulfill clinically significant demands throughout the device's life span.
- *Implementation of Good Software Engineering and Security Practices:* The model is designed with the "fundamentals" in mind: excellent software engineering methods, data quality assurance, data management, and effective cybersecurity policies. Methodical risk management and design processes that can document and convey design, implementation, and risk management decisions and reasons, as well as ensure data authenticity and integrity, are examples of these techniques.

- *Clinical Study Participants and Datasets Represent the Targeted Patient Population:* Data collection protocols should ensure that the relevant characteristics of the intended patient population (such as age, gender, sex, race, and ethnicity), use, and measurement inputs are adequately represented in a sample of sufficient size in the clinical study and training and test datasets so that results can be reasonably generalized to the population of interest. This is necessary in order to control any bias, promote suitable and generalizable performance across the intended patient group, test usability, and identify situations in which the model may underperform.

- *Training Data Sets Are Independent of Test Sets:* Training and test datasets are chosen and maintained to be properly independent of one another. All possible sources of reliance, such as patient, data gathering, and site characteristics, are assessed and handled to ensure independence.

- *The Best Available Methods Are Used to Select Reference Datasets:* Accepted, best-available procedures for establishing a reference dataset (that is, a reference standard) ensure that clinically relevant and well-characterized data are collected and that the reference's limits are known. Accepted reference datasets, if available, are utilized in model development and testing to promote and show model robustness and generalizability across the intended patient population.

- *Model Design Is Tailored to the Available Data and Reflects the Device's Intended Use:* Model design is appropriate for the given data and allows for the active mitigation of recognized hazards such as overfitting, performance degradation, and security threats. The clinical advantages and hazards associated with the product are well recognized, utilized to establish clinically valid performance goals for testing, and support that the product may safely and effectively fulfill its intended purpose. Considerations include the influence of both global and local performance, as well as uncertainty/variability in device inputs, outputs, targeted patient demographics, and clinical usage settings.

- *Emphasis on the Performance of the Human–AI Team:* Where the model has a "human in the loop," human factors issues and the human interpretability of model outputs are handled with an emphasis on the performance of the human–AI team rather than the performance of the model in isolation.

- *Testing Demonstrates Device Performance under Clinically Relevant Conditions:* Statistically sound test plans are devised and implemented to yield clinically relevant device performance information independent of the training dataset. Considerations include the desired patient population, relevant subgroups, the clinical setting and utilization by the human–AI team, measurement inputs, and any confounding variables.

- *Users are Given Concise, Essential Information:* Users have ready access to clear, contextually relevant information that is appropriate for the intended audience (such as healthcare providers or patients), such as the product's intended use and indications for use, the model's performance for appropriate subgroups, the characteristics of the data used to train and test the model, acceptable inputs, known limitations, user interface interpretation, and clinical workflow integration. Users are also made aware of device alterations and upgrades as a result of real-world performance monitoring, which serves as a foundation for decision-making when available and a channel for communicating product problems to the developer.

- *Performance of Deployed Models Is Monitored, and Retraining Risks Are Managed:* Deployed models can be monitored in the "real-world" use to maintain or improve safety and performance. Moreover, when models are trained on a standard or continuous basis after deployment, appropriate controls are in place to manage risks of overfitting, unintended bias, and model degradation (for example, dataset drift), all of which may have an impact on the model's safety and performance as it is used by the human–AI team (50).

10.10 CONCLUSION

The incorporation of AI and ML solutions in the health sector offers enormous potential to close inequality gaps, rehumanize medicine, and enhance people's lives. However, the publicity surrounding this integration must be controlled properly. The ethics of this merger must be thoroughly explored if medicine is to remain a human-centered enterprise. As a result, models must be created with the end user in mind.

These solutions must take into account the impact on informed consent, safety and transparency, algorithmic fairness and biases, and data privacy. People will be eager to accept these technologies if they are given due attention and hype management. Aside from local governments and citizens, stakeholders from around the world, such as the World Health Organization (WHO), the Council of Europe, the Organization for Economic Co-operation and Development (OECD), and the United Nations Educational, Scientific, and Cultural Organization (UNESCO), should collaborate to develop a common plan to address the ethical challenges and opportunities of using AI for health, such as through the United Nations Interagency Committee on Bioethics, so that these technologies can flourish and revolutionize the world of medicine.

REFERENCES

1. World Economic Forum. 2017. The age of robots could be a new Renaissance. This is why. [accessed 2022 October 28]. https://www.weforum.org/agenda/2017/10/ai-renaissance/.
2. IFPMA. 2018. Present to future: Trends and challenges. Technology: Can it deliver a more precise, individual and empowering approach to health? [accessed 2022 October 28]. https://50years.ifpma.org/presenttofuture/technology/.
3. Topol EJ. 2019. High-performance medicine: The convergence of human and artificial intelligence. Nat Med. 25:44–56. [accessed 2022 October 28]. https://www.nature.com/articles/s41591-018-0300-7.
4. Noor P. 2020. Can we trust AI not to further embed racial bias and prejudice? BMJ Feb 12;368:m363. [accessed 2022 October 28]. https://www.bmj.com/content/368/bmj.m363.long.
5. Morley J, Machado CCV, Burr C, Cowls J, et al. 2020. The ethics of AI in health care: A mapping review. Soc Sci Med. Sep;260:113172. [accessed 2022 October 28]. https://www.sciencedirect.com/science/article/abs/pii/S0277953620303919?via%3Dihub.
6. US Food and Drug Administration. 2021. Artificial Intelligence/machine learning (AI/ML)-based Software as a Medical Device (SaMD) action plan. [accessed 2022 October 29]. https://www.fda.gov/medical-devices/software-medical-device-samd/artificial-intelligence-and-machine-learning-software-medical-device.
7. Calo R. 2018. Artificial intelligence policy: A primer and roadmap. University of Bologna Law Review. 3(2):180–218. [accessed 2022 October 29]. https://bolognalawreview.unibo.it/article/view/8670/8420
8. Reddy S, Allan S, Coghlan S, Cooper P. 2020. A governance model for the application of AI in health care. J Am Med Inform Assoc. 27(3):491–497. [accessed 2022 October 29]. https://www.ncbi.nlm.nih.gov/pmc/articles/PMC7647243/.
9. Cerys Wyn Davies. 2021. Why healthcare providers need a policy on AI ethics. OUT-LAW ANALYSIS. [accessed 2022 October 30]. https://www.pinsentmasons.com/out-law/analysis/why-healthcare-providers-need-a-policy-on-ai-ethics.
10. Report of the Special Rapporteur on the promotion and protection of the right to freedom of opinion and expression. Artificial Intelligence and its impact on human rights (A/73/348) Online content regulation (A/HRC/38/35) Encryption and Anonymity – follow up report (A/HRC/38/35/Add.5). [accessed 2022 November 02]. https://freedex.org/wp-content/blogs.dir/2015/files/2018/10/AI-and-FOE-GA.pdf.
11. Organization for Economic Co-operation and Development. 2019. Recommendation of the Council on Artificial Intelligence (OECD Legal Instruments. OECD/LEGAL/O449). Paris. [accessed 2022 November 02]. https://legalinstruments.oecd.org/en/instruments/OECD-LEGAL-0449
12. Hao K. What is machine learning? Machine-learning algorithms find and apply patterns in data. And they pretty much run the world. MIT Technology Reviews, November 17, 2017. [accessed 2022 November 02]. https://www.technologyreview.com/2018/11/17/103781/what-is-machine-learning-we-drew-you-another-flowchart/.

13. Vinuesa R, Azizpour H, Leite I, et al. 2020. The role of artificial intelligence in achieving the sustainable development goals. Nat Commun.11:233. [accessed 2022 November 02]. https://www.nature.com/articles/s41467-019-14108-y.

14. Konieczny L, Roterman I. 2019. Personalized precision medicine. Bio-Algorithms and Med-Systems. 15(4):20190047. [accessed 2022 November 03]. doi: 10.1515/bams-2019-0047.

15. Erin PB, Bryan TM, John RB. 2015. National Academies of Sciences, Engineering, and Medicine: Improving Diagnosis in Health Care. Washington, DC: The National Academies Press. [accessed 2022 November 03]. https://nap.nationalacademies.org/read/21794/chapter/7#243.

16. Bohr A, Memarzadeh K. 2020. The rise of artificial intelligence in healthcare applications. Artificial Intelligence in Healthcare. 2020:25–60. [accessed 2022 November 04]. https://www.ncbi.nlm.nih.gov/pmc/articles/PMC7325854/.

17. Velupillai S, Suominen H, Liakata M, et al. 2018. (2018). Using clinical natural language processing for health outcomes research: Overview and actionable suggestions for future advances. J Biomed Inform. 88:11–19. [accessed 2022 November 04]. https://www.sciencedirect.com/science/article/pii/S1532046418302016?via%3Dihub

18. Hoyt RE, Snider D, Thompson C, Mantravadi S. 2016. IBM Watson analytics: Automating visualization, descriptive, and predictive statistics. JMIR Public Health and Surveillance. 2(2):e157. [accessed 2022 November 04]. https://www.ncbi.nlm.nih.gov/pmc/articles/PMC5080525

19. Powles J, Hodson H. 2017. Google DeepMind and healthcare in an age of algorithms. Health Technol. 7(4):351–367. [accessed 2022 November 04]. https://link.springer.com/article/10.1007/s12553-017-0179-1

20. AI in India: A Policy Agenda—The Centre for Internet and Society. [accessed 2022 November 04]. https://cis-india.org/internet-governance/blog/ai-in-india-a-policy-agenda.

21. Shaikh J, Suganda Devi P, Shaikh MA, et al. 2020. Role of artificial intelligence in prevention and detection of Covid-19. Int J Adv Sci Technol. 29:45–54. [accessed 2022 November 04]. http://sersc.org/journals/index.php/IJAST/article/view/13000.

22. Denecke K, Abd-Alrazaq A, Househ M. 2021. Artificial Intelligence for Chatbots in Mental Health: Opportunities and Challenges. In: Househ, M., Borycki, E., Kushniruk, A. editors. Multiple Perspectives on Artificial Intelligence in Healthcare. Lecture Notes in Bioengineering. Springer, Cham. pp. 115–128. [accessed 2022 November 04]. https://www.researchgate.net/publication/353726195_Artificial_Intelligence_for_Chatbots_in_Mental_Health_Opportunities_and_Challenges.

23. Cabestany J, López CP, Sama A, Moreno JM, Bayes A, Rodriguez-Molinero A. 2013. REMPARK: When AI and technology meet Parkinson Disease assessment. Proceedings of the 20th International Conference Mixed Design of Integrated Circuits and Systems; June 20–22, 2013, Gdynia, Poland. MIXDES 2013. p. 148. Department of Microelectronics & Computer Science, Lodz University of Technology, Poland. [accessed 2022 November 04]. https://libstore.ugent.be/fulltxt/RUG01/002/047/809/RUG01-002047809_2013_0002_AC.pdf

24. Paul D, Sanap G, Shenoy S, Kalyane D. 2021. Artificial intelligence in drug discovery and development. Drug Discov Today. 26(1):80–93. [accessed 2022 November 04]. https://www.ncbi.nlm.nih.gov/pmc/articles/PMC7577280/

25. Khanna V, Ahuja R, Popli H. 2020. Role of artificial intelligence in pharmaceutical marketing: A comprehensive review. J Adv Sci Res. 11(03):54–61. [accessed 2022 November 05]. https://sciensage.info/index.php/JASR/article/view/506/226.

26. Babel A, Taneja R, Mondello MF. 2021. Artificial intelligence solutions to increase medication adherence in patients with non-communicable diseases. Front Digit Health. 2021 Jun 29;3:669869. [accessed 2022 November 06]. https://www.ncbi.nlm.nih.gov/pmc/articles/PMC8521858/.

27. Davenport T, Kalakota R. 2019. The potential for artificial intelligence in healthcare. Future Healthc J. 6(2):94–98. [accessed 2022 November 6]. https://www.ncbi.nlm.nih.gov/pmc/articles/PMC6616181/

28. Baweja S, Singh S. 2020. "Beginning of Artificial Intelligence, End of Human Rights." [accessed 2022 November 06]. https://blogs.lse.ac.uk/humanrights/2020/07/16/beginning-of-artificial-intelligence-end-of-human-rights/.

29. Council of Europe. 2019. Guidelines on artificial intelligence and data protection. Consultative committee of the convention for the protection of individuals with regard to automatic processing of personal data. Jan 25, Strasbourg. [accessed 2022 November 07]. https://rm.coe.int/guidelines-on-artificial-intelligence-and-data-protection/168091f9d8.

30. Council of Europe. 2018. European ethical charter on the use of artificial intelligence in judicial systems and their environment. 31st plenary meeting of the CEPEJ. December 3–4, Strasbourg. [accessed 2022 November 07]. https://rm.coe.int/ethical-charter-en-for-publication-4-december-2018/16808f699c.

31. Edward SD. 2018. The EU General Data Protection Regulation: Implications for international scientific research in the digital era. The Journal of Law Medicine & Ethics. 46(4):1013–1030. [accessed 2022 November 07]. https://www.researchgate.net/publication/330316678_The_EU_General_Data_Protection_Regulation_Implications_for_International_Scientific_Research_in_the_Digital_Era

32. Edemekong PF, Annamaraju P, Haydel MJ. Health Insurance Portability and Accountability Act. [Updated 2022 Feb 3]. In: Stat Pearls [Internet]. Treasure Island (FL): Stat Pearls Publishing; 2022 Jan. [accessed 2022 November 07]. https://www.ncbi.nlm.nih.gov/books/NBK500019/.

33. European Union. 2019. General recommendations for the processing of personal data in artificial intelligence. Brussels: Red IberoAmerica de Proteccion de Datos. [accessed 2022 November 07]. https://www.redipd.org/sites/default/files/2020-02/guide-general-recommendations-processing-personal-data-ai.pdf.

34. European Union. 2019. Specific guidelines for compliance with the principles and rights that govern the protection of personal data in artificial intelligence projects. Brussels: Red IberoAmerica de Proteccion de Datos. [accessed 2022 November 07]. https://www.redipd.org/sites/default/files/2020-02/guide-specific-guidelines-ai-projects.pdf.

35. World Health Organization. 2021. Ethics and Governance of Artificial Intelligence for Health: WHO Guidance. Geneva. Licence: CC BY-NC-SA 3.0 IGO. [accessed 2022 November 07].

36. What Is Human in the Loop (HITL) Machine Learning? – BMC Software | Blogs. [accessed 2022 November 07]. https://www.bmc.com/blogs/hitl-human-in-the-loop/.

37. Gerke S, Minssen T, Cohen G. 2020. Ethical and legal challenges of artificial intelligence-driven healthcare. Artificial Intelligence in Healthcare, 295–336. [accessed 2022 November 07]. https://www.ncbi.nlm.nih.gov/pmc/articles/PMC7332220/.

38. Linda AM. 2022 March 17. Data privacy and artificial intelligence in health care. Reuters. [accessed 2022 November 10]. https://www.reuters.com/legal/litigation/data-privacy-artificial-intelligence-health-care-2022-03-17/

39. Council of Europe. 2021. Guide to Public Debate on Human Rights and Biomedicine. 19–21 November Strasbourg. [accessed 2022 November 10]. https://rm.coe.int/prems-009521-ex-061320-gbr-2007-guide-on-public-debate-16x24-web/1680a12679.

40. Panch T, Mattie H, Celi LA. 2019. The "inconvenient truth" about AI in healthcare. NPJ Digit Med. 2, 77. [accessed 2022 November 11]. https://www.nature.com/articles/s41746-019-0155-4.

41. Cruz RZ, Liu X, Chan AW, et al. 2020. Guidelines for clinical trial protocols for interventions involving artificial intelligence: The SPIRIT-AI extension. Nat Med. 26(9):1351–1363. [accessed 2022 November 11]. https://www.nature.com/articles/s41591-020-1037-7.

42. Koul A, Bawa RK, Kumar Y. 2022. Artificial intelligence techniques to predict the airway disorders illness: A systematic review. Arch Computat Methods Eng. [accessed 2022 November 12]. https://link.springer.com/article/10.1007/s11831-022-09818-4#citeas.

43. Wahengbam K, Singh MP, Nongmeikapam K, et al. 2021. A group decision optimization analogy-based deep learning architecture for multiclass pathology classification in a voice signal. IEEE Sens. J. 21(6):8100–8116. [accessed 2022 November 11]. https://ieeexplore.ieee.org/document/9314040.

44. Gaube S, Suresh H, Raue M, et al. 2021. Do as AI say: Susceptibility in deployment of clinical decision-aids. NPJ Digit. Med. 4. [accessed 2022 November 11]. https://www.nature.com/articles/s41746-021-00385-9

45. Pchelintseva OI, Omelyanskaya OV. 2022. Features of conducting ethical review of research on artificial intelligence systems on the basis of the research and practical clinical center for diagnostics and telemedicine technologies of the Moscow health care department, Moscow, Russian. Digital Diagnostics. 3(2):156–161. [accessed 2022 November 12]. https://jdigitaldiagnostics.com/DD/article/view/107983.

46. Schiff D, Borenstein J. 2019. Case and commentary: How should clinicians communicate with patients about the roles of artificially intelligent team members? AMA J Ethics. 21(2):E138. [accessed 2022 November 12]. https://journalofethics.ama-assn.org/article/how-should-clinicians-communicatepatients-about-roles-artificially-intelligent-team-members/2019-02

47. US Food and Drug Administration. 2021. Discussion Paper: Proposed Regulatory Framework for Modifications to Artificial Intelligence/Machine Learning (AI/ML)-based Software as a Medical Device (SaMD). [accessed 2022 November 12]. https://www.fda.gov/medical-devices/software-medical-device-samd/artificial-intelligence-and-machine-learning-software-medical-device.

48. US Food and Drug Administration. 2020. Executive Summary for the Patient Engagement Advisory Committee Meeting. Oct 22. [accessed 2022 November 12]. https://www.fda.gov/advisory-committees/advisory-committee-calendar/october-22-2020-patient-engagement-advisory-committee-meeting-announcement-10222020-10222020.

49. US Food and Drug Administration. 2017. What Are Examples of Software as a Medical Device? Software as a Medical Device? [accessed 2022 November 12]. https://www.fda.gov/medical-devices/software-medical-device-samd/what-are-examples-software-medical-device#:~:text=Software%20as%20a%20Medical%20Device%20ranges%20from%20software%20that%20allows,to%20help%20detect%20breast%20cancer.

50. US Food and Drug Administration. 2021. Good Machine Learning Practice for Medical Device Development: Guiding Principles USFDA. [accessed 2022 November 12]. https://www.fda.gov/medical-devices/software-medical-device-samd/good-machine-learning-practice-medical-device-development-guiding-principles.

11 Role of Artificial Intelligence and Machine Learning in Nanosafety

Vijaya Rajendran
Anna University, Chennai, Tamil Nadu, India

Subha Sri Ramakrishnan
Anna University, Chennai, Tamil Nadu, India

Ronaldo Anuf A.
Kamaraj College of Engineering and Technology,
Vellakulam, Tamil Nadu, India

Kiruba Mohandoss
Sri Ramachandra Institute of Higher Education and
Research, Chennai, Tamil Nadu, India

11.1 INTRODUCTION

Nanotechnology is acknowledged as an authorized technology reinforcing strengthening astride over industrial, governmental, and societal sectors as their implementations are safe and conservative. Globally, a profound fraction of total investment is enthralled on the safety characteristics of nanotechnology due to the occupational revelation and consumer asylum as their priority areas. Nanotechnology is vital as they operate on small-size scale which equipage protein and other biomolecules harnessed in nature to impart cellular structure and function by coalescing them to their surfaces and equipping them with competently fascinating cellular receptors. Nanomaterials (NMs) have unparalleled contact to living systems, contributing the foreseeable to decussate biological barriers to earmark definite cells and to dispatch cargo to particular organelles. Nanotechnology underpins its dominance beyond industrial, governmental, and societal sectors as NMs hold entrancing process; besides, safety aspects of NMs from occupational exposure and consumer safety to environmental impacts are priority areas in scientific community (1, 2).

Nanosafety is a cornerstone of superintendence with exhilarating area of research progresses in cheminformatics, toxicology, small-molecule–ligand chemistry and amalgamating them in analytical tactics to turnout exhilarating novelties (3). Nanosafety evaluation is supreme as it is obligatory not only for human health preservation and environmental integrity but also as a keystone for industrial and regulatory bodies (4). Nanosafety persists safety evaluation over the product life cycle by improving health and environmental impacts under rational exposure and vulnerability constraints. New approach methods (NAMs) foster shielding and safer designing of nanomaterials with enabled automations (5).

Robust endeavor to perorate safer materials with diminished ambiguity satisfactorily authorizes mercantile acquisition of nano-enabled technologies. This entails future orientation with regard to safety assessment of more new-fangled matters. Modern nanosafety research evaluates the biological

DOI: 10.1201/9781003343981-11

impacts of cardinal NMs properties such as size, solubility, charge, and unequivocal proportions shape (6).

11.2 LIFE CYCLE OF NM

NMs are innate or engineered materials grounded on nanosized particles in a disarticulated state or in the shape of a cluster/agglomerate. Nanoparticles (NPs) possess unique physiochemical and structural properties as they are extensively employed in the drug applications. The life cycle (Figure 11.1) of NM is recognized as a significant tool for evaluating the potency of the particle toward the environment impacts in their manufacturing during their complete life cycle. The unique properties of the NMs are their higher surface area-to-volume ratio of plasmonic properties.

Engineered nanoparticles (ENPs) have versatile applications due to their inherent properties. Surface functionalization is the key parametric in engineering NPs.

11.3 SAFETY PROFILE IN DESIGNING NANOMEDICINE

Nano-objects, NPs, and their agglomeration and aggregation are the major prospectives in the industrial innovation in a variety of fields. Besides their advancement in the drug delivery and versatile applications, some exert toxic nature to the environment and the biota due to their properties and exposures. ENPs constitute major toxicity in biological life due to their inherent functionalization and their surface groups which result in materialistic interactions in the biota (7). Animal studies buoyed that the exposure to the NMs has harmful effects on the human and environment. Ensuring safety of nano-enabled products is a crucial element besides the potential advantages of the NMs and their inherent benefits (8). Risk assessment (RA) of ENM has substantial limitations in supporting the health impact assessment of this new class of chemicals. There are various regulatory bodies established to govern the safety profile of NMs. *Nanosafety Cluster—* an EU Commission initiative released "Strategic Research Agenda toward Safe and Sustainable Nanomaterial and Nanotechnology Innovations" for enhancing, accelerating, and addressing all

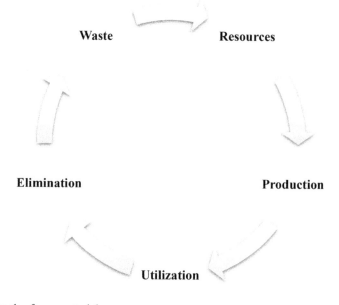

FIGURE 11.1 Lifecycle of nanomaterials.

aspects of nanosafety (9). Screening strategies entails *in vitro* and *in vivo* assays which has been advanced for accelerating the degree of hazard prediction with high-throughput and high-content screening laterally with alternative test strategies as a powerful tool to expedite the stride of hazard detection and RA. These data advance safety assessment with the intricate and multifaceted nature of events stirring at the nano–bio interfaces or boundaries with the complete replacement of *in vivo* assessment. The nature of this interface is multifaceted, changeable, and the biological behavior is unpredictable solely on the basis of size, shape, and surface chemistry as these intrinsic properties are exposed to alteration in response to the biological environment (8). The issue of nano-specificity of biomarkers is challenging due to the reduction in size of the particles to nanoscale and can drive the new and unpredictable properties of the certain NPs which is apparent and may differ from the properties exhibited in the bulk counterpart.

Before heading a drug toward market, the safety profile should be monitored as it is crucial as it interrupts the biota. REACH (Registration, Evaluation, Authorization, and Restriction of Chemical Substances) is implicit in safety profile assessment of NMs and the risk assessment methodology being adopted for the conventional therapeutics. Risk assessment of NMs constitute three protocols:

a. Evaluation of effects
b. Routes of exposure and their assessments
c. Risk characterization

Curatively, the evaluation of possible NM-related risks is a luxurious and byzantine errand and accomplished by the integration of *in vitro* and *in vivo* experiments which elucidates the potent biota and environmental hazards associated with NMs. The studies in analyzing the toxicity parameters of NMs is complex due to versatility of structural properties exhibited with aggregation and agglomeration. NMs safety is a preeminent bottleneck in the disbandment of innovative nano-enabled products to the market. To address these challenges, predictive *in silico* approaches are taken into account for analyzing the intricacy of NMs and their varied environmental hazards associated with functionalizing them and it is an essential part of nanosafety research (10).

11.4   DEFINITIVE STEP ACROSS COMPUTATIONAL METHODS

Computational methodologies are prevailing as an alternative carrier in contrast to the traditional *in vivo* tests for the regulatory applications of QSARs. It will be used progressively as accessible and reliable models. OECD proposed five principles for the computational techniques in the regulatory context:

a. A distinct endpoint
b. An unequivocal algorithm
c. A characteristic demesne of pertinence
d. Pertinent proportion of goodness of fit, robustness, and prophecy
e. An automatous explication

The supreme efficacious models having potential for predicting biological properties of NMs in a sundry and intricate backgrounds are based on the quantitative structure–activity relationship (QSAR) method. These methods employ statistical and machine learning (ML) algorithms and prototypical interrelation among the materials, structure, molecular properties, derivation, and other constraints; based on their applications, some of the tools are scrutinized with their potent biological effects. The persuasive hindrance of these QSAR models is that they employ small datasets and large integrated data can't be optimized by these methods (11).

AI-based computational methodologies forecast toxicological properties of chemical complexes due to the exhaustive novelties instigated in computational methods. The integration of AI and ML learning techniques paved way for the molecules in reducing toxicity profile and ensuring safety in NPs (12).

AI techniques are segregated as weak and strong based on their function. Weak AI is the one which accomplish very explicit tasks and programs with excellent exactness, which is in contrast to strong AI where it has the skill to pursue in accordance with the way humans do. In ML techniques, tools are resolved as supervised learning and unsupervised learning. Supervised learning is a set of input and output data pairs required for training; whereas the input data are furnished in the raw form for the algorithms in unsupervised learning, which in turn try to explore approximate beneficial aspect in the specified data. Contrary to traditional ML, reinforcement learning techniques custom algorithms that intermingle with the environment persistency tasks sequentially while rectifying actions to intensify consummation.

Specifically, artificial neural networks (ANNs) custom algorithms that master through contemplation of instruction form data in a categorized layer with nonlinear refining units. Deep neural network (DNN), denoted as deep learning (DL), is a riotous technology which can be successfully passed on to a wide range of multiplex contexts such as forecasting toxicities for chemicals by analyzing physiochemical and biophysical properties of recognized substances from the millions of resources [13]. Some methods used in toxicology in DL include deep convolutional neural networks (CNNs) and self-normalizing neural networks.

DL paradigms acquire significantly huge data for equipping and additional computing power to hold all that data, approximately which are inadequate methodologies to date. In recent years, the neural network-based methodologies and their novel computational method in the realm of ML and DL lead to the popularized factors due to their rapid development and commercialization due to the free accessibility of the databases in the field of medical, chemical, and pharmacological knowledge and electronic medical records.

The future perspective of this computational technologies employs large integrated datasets as they upsurge the automation and mechanization of the experimental and investigational data analysis and interpretation. These methods can promptly block data gaps and exploit "read across" extrapolation of biological effects of the similar materials which in turn categorize the hazards of NMs to distinct species.

The most recent breakthrough in nanoinformatics domain embraces five important fields and also their assessment, expansion, exploration, incorporation and integration in the future generations of NM computational modeling tools and infrastructures:

   i. Datasets curation, quality assessment, and knowledge infrastructure
  ii. Toxic genomics modeling
 iii. Multiscale modeling (physics-based and data-driven)
  iv. Predictive modeling (data-driven)
   v. NMs human and environmental RA

These advances are exploited within the European Commission Horizon 2020 funded project NanoSolveIT which create relevant nanoinformatics e-platform to simplify *in silico* NMs exposure, hazard, and risk assessment.

11.5 TOXICITY DUE TO DDIS

Computational modeling decodes the toxicological interactions with a correlation with biokinetic and dynamic models with an integration and evolution of artificial intelligence (AI) and ML. Frequently, biology-based mathematical models such as Bayesian methods and Markov chain Monte Carlo simulation are highly reliable for the experimental systems of nanotherapeutics and engineering nanomaterials (ENMs). ENMs are more prone to the toxicity as they possess inherent physiochemical and structural properties (14).

Physiochemical Property of Nanomaterials	Influence
Plasmonic property	Strong interaction with incident light and influence biological characteristics of a particle
Gap band factor	• Impact in optical and redox properties, risk of generating ROS • Enthalpy becomes negative as the energy of conducting band NP hydrates • Overlap of NM and cellular conductance bands facilitates transfer of electrons and oxidative stress
Surface coating	• Affects surface load and assists in stability of particle • Recognizes and reacts with specific bonds • Reactivity changes occur
Particle size	
Surface electrical charge	Interaction with subsystems or biological membranes and damages cell membrane
Ionic dissolution	Higher ionic dissolution leads to higher toxicity
Shape	Anisotropic or rod-shaped NMs are less efficient
Crystallinity	Amorphous forms are less toxic when compared with the other quartz silica and asbestos
Surface bond strain	High curvature and high temperature synthesis exerts high toxicity
Dissolution	Biopersistence and bio durability
Formation of biomolecular corona	Characterization

11.6 COMPUTATIONAL NANOSAFETY ROADBLOCKS, MILESTONES

European Cooperation in Science and Technology progresses the use of ML methods in nanosafety research and design of effective and safe NMs for the commercial applications. *In vivo* studies in the laboratory animals which are paradoxically prognostic for the human and pathophysiology are problematic as they encompass animal ethical concerns. *In vitro* assays that correlate with *in vivo* results are essential as they afford precise info and ideas and aid in the modeling of ML-based *in vivo* models. There are versatile tools being engineered and employed for quantifying the toxicity profile of the NM such as ToxML (toxicology database language), ACToR which is an online warehouse of all publicly obtainable chemical toxicity data, and expansion of ontologies which is a formal demonstration of knowledge as a set of concepts, and the relationships between those concepts for NMs.

Initially, a five-year time frame implicit a potent increase in the capacity of *in vivo* data on the effect of NPs, advancement of data storage and sharing methods, progression of reliable models for *in vivo* and *in vitro* models with the generation of first models of NPs with different environments, elucidation of mechanism in their route of exposure to the cells, and exerting the level of toxicity such as free-radical production, apoptosis, reactive oxygen species, genotoxicity, cytotoxicity, and so on.

Next milestone was exhibited in the next ten years, that is, the manufacture of ML models of *in vitro* and *in vivo* sequel of NPs satisfactorily authentic for regulatory purposes, burgeoning of models that can valid forecast NP corona in variegated environments, and progression of NMs cataloging fingerprints, that is, the physicochemical, genomic, and/or biological profiles of NMs which cluster materials with parallel *in vivo* effects and permit regulators to categorize NMs into hazard modules in an analogous protocol which are currently implemented for industrial chemicals.

Accomplishing these milestones in this roadmap is indispensable in the perpetuation of a network of hypothetical and computational researchers, controllers, regulators, investigators, and supervisory. It is done through a series of EU Projects and Movements (MODENA, MARINA, NANOSOLUTIONS) more recently by analyzing and funding of at least three EU Horizon 2020 projects: NanoSolveIT, SABYDOMA, and Nano Commons. These milestones also presumed that high-throughput and qualitative investigation procedures for construction and characterization of NMs would ascend in furnishing data for equipping ML models for accomplishing this

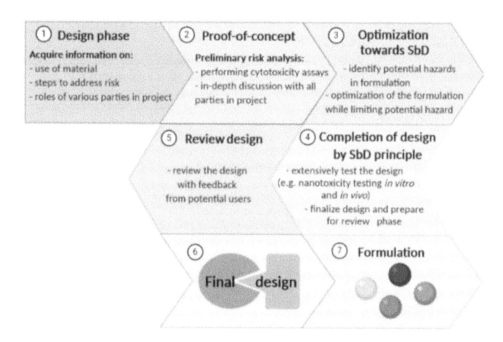

FIGURE 11.2 Phases in risk assessment of nanoformulation.

pioneering sets of milestones in the progression of NMs for both practical and coherent custom in accordance with safe by design principle. Figure 11.2 shows the phases in risk assessment of nanoformulation.

11.7 UNRESOLVED ROADBLOCKS HINDERING THE EVOLUTION OF ML IN NANOSAFETY

In guaranteeing seven years, these striving milestones have been accomplished incompletely because the automatic NMs synthesis and characterization methods have not been implemented due to the restraining of the accessibility of data for training models. A new set of nanoinformatics milestones for 2030 is lately distinct essentially echoing, extending, elaborating, and summarizing the state-of-the-art diverse research areas relevant to NM risk assessment and governance.

ML methods are reliant on satisfactory data for the training and authentication and on the cohort of the relevant descriptors which characterizes the properties of NMs which is reliant on the procuration of the valuable subset of the descriptors and robust training of models and its endorsement of the extrapolative power of models and the commissioning of it to forecast the properties of the new and upgraded materials.

The foremost crucial thing in the descriptor generation is the selection and authentication of the good descriptor with ML algorithm that will engender a useful prototype, whereas descriptors which are poor representatives of materials will spawn very poor models (15).

11.8 NANO-QSAR CYTOTOXICITY MODELS

Standard QSAR models can't be adapted for the computational screening of NMs and so different models are being made available due to the inherent limitations. A computational hybrid nano-QSAR model for the nanocytotoxicity approves twofold descriptors: enthalpy of a development (associated with bandgap energy) and electronegativity (correlated to stability). It also predicts biopersistence of ENMs which aids in the safety protocols. In order to foster the progression of computational tools for the assessment and level of toxicity persuaded by ENMs, European Commission has launched different schemes and protocols under REACH regulation. Based on analyses drawn from the nano-cluster growth rate and physicochemical properties of the ENMs, a read across of the toxicological properties ENMs based on the governance, risk, and compliance conversion can be accomplished.

11.9 PAUCITY OF DATASETS TO TRAIN ML MODELS

ML techniques are highly reliant on the quantitative and qualitative aspects of diverse datasets hired to train the models which is more consistent and can forecast the properties of new materials that is not used to train the models. The main limitations are that the unavailability and precise data employed for the descriptors do not contain sufficient information about the molecular, physiochemical, and structural characteristics of the NMs to spawn a robust and predictive model. Now researchers are employing the read across model for predicting the nanosafety assessment and evaluation. This model is non-experimental method for filling data gaps relayed on the properties of close analogue or similar chemical category.

Figure 11.3 depicts the roles of AI techniques.

11.10 APPLICATION OF AI AND ML TO NANOSAFETY

The integration of ML or statistical modeling to predict the adverse properties of NMs was elucidated by Puzyn et al. (16). They discovered a simple linear regression model which predicted the cytotoxicity of 17 diverse metal oxide NPs to *Escherichia coli* using the descriptors resultant from the quantum chemical calculations. Epa et al. (17) described the custom of linear regression and Bayesian standardized neural networks to forecast the biological outcome of 51 metal oxide NPs with assorted metal cores and 109 metal oxide NPs with parallel cores but varied surface modifiers. Weissleder et al. reported the cellular uptake of 109 NPs with a core of super magnetic NPs and dextran coating and stands versatile small molecules conjugated to their surface and provoked nano-QSAR models for the cellular uptake. Two cell lines pancreatic cancer (PaCa2) and human umbilical vein endothelial cell (HUVEC) exhibited potent variation in the ingestion of the surface-modified NP (18).

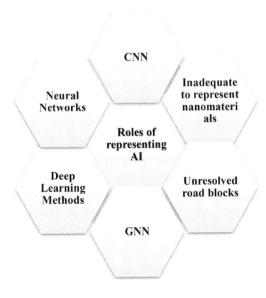

FIGURE 11.3 Different roles of artificial intelligence.

11.11 NANOMATERIALS DATASETS

Nanotoxicity prediction is vital for exploring risk assessment of ENP and its routes of exposure and a critical factor is assessing the safety profile of NMs. The computational models in forecasting the nanosafety of NMs are crucial and entail the accessibility of good datasets with high data superiority, extensiveness, and quantity which is restrained in terms of number of varied NMs and also integrating a wide range of doses, timepoints, cell or organism types, and endpoints. The formulating industry of the NMs are accountable for the safety assessment profile and its exposure and potent risk assessment associated with synthesizing it and also the provision of data for their specific NMs for regulatory evaluation research. Comprehensive research is obligatory for the amplification of anticipating models to authorize curtailment of the experimental and investigational data requisites for regulatory RA. Computational researchers forehead momentous threads in receiving this data desired for the progression of robust models. ENPs entail intense surface functionalization, characterization, routes of exposure to the environment, and hazardous data needed to be provided as a dataset for the computational modeling for *in vitro* and *in vivo* studies. These underpinning datasets are required for modeling and employ capricious degree of completeness and intense quality. This intense research is focused on edifying the datasets for feeding in assessing the safety profile of an NP. In the field of nanoinformatics, special prominence is focused on the curation and quality assessment of nanosafety data. A trivial number of new projects have thus prioritized the curation of literature data and datasets generated in past project that ended before the current (evolving) standards for data quality emerged (NanoReg2 and caLIBRAte projects curating NaNoReg data and NanoSolveIT partners curating literature publications on NMs transcriptomics) (19).

A huge number of experimental approaches have been published which led to some of the missing values and differing data quality like number of replicates, relevance of endpoints, and dissimilar experimental conditions, so the quality assessment and data completion are serious issues for regulators, modelers, controllers, and supervisors (20). They anticipated standards to evaluate the consistency of toxicological and eco-toxicological data based on the source of the toxicity data whether the given data were fashioned using international standard operating procedures (SOPs) such as the above-mentioned OECD test guidelines. These lengthened standards comprise PChem properties of NMs such as surface charge, size, shape, and surface functionalization. These further protracted criteria for assessment of the quality and completeness of NMs PChem data.

11.12 EXAMPLES OF SAFETY-BY-DESIGN OF NANOMODELING

Feng et al. forecasted the safety of metal oxide NPs which possess nifty applications and potential toxicological injury. The safe by design strategy is governed to regulate the physiochemical properties of the NPs by NM edge energy. Manganese oxide (Mn_3O_4) NPs, toxic metal NPs, are being dopped with various transition metals for the purpose of conduction band. Fermi energy (E_F) moves zinc (Zn)-, copper (Cu)-, and chromium (Cr)- dopants far from the valence band. *In vitro* and *in vivo* studies assessments reveal that Zn-, Cu-, and Cr-doped Mn_3O_4 NP could generate lower water groups and weaken the injury in contrast to the Mn_3O_4 NPs which demonstrate lower toxicity profile (21).

Georgia Melagraki proposed that zeta potential is the critical property of an NP which provides the estimation of the surface charge, electronic stability, and also the tendency to aggregate as agglomerates and interact with the cellular membranes. He utilized the input data as set of image descriptors derived from the TEM and a tool NanoXtract which in turn validated the NM behavior and its biological effects, core composition, and similarity features with the other particles, which is a key parameter for predicting the toxicity profile of NMs. He also employed these computer descriptors to refashion a read across model for the extrapolation of zeta potential based on the k-nearest neighbors approach and this execution aided for gap-filling of datasets and expedite the encroachment of prognostic and *in silico* nanosafety assessment.

An exciting feature of NM is its interior bio-corona and Li Shang et al. familiarize a probe protein corona formation utilizing fluorescence resonance energy *in situ* on quantum dots. He utilized an absorption of human serum albumin (HSA) onto the surface of InP@ZnS quantum dots with dissimilar chirality, which also interprets the strong apparent variances in the binding behaviors comprising affinity and adsorption orientation that are acquired upon quantitative analysis. Circular dichroism spectroscopy auxiliary authorizes the variances in the conformational changes of HSA upon collaboration with D- and L-chiral QD surfaces. Subsequently, the fashioned protein corona on chiral surfaces may distress their consecutive biological interactions, protein exchange with serum proteins plasma, as well as cellular interactions. These results intensely demonstrate the probable of the FRET technique as an unpretentious yet adaptable podium for quantitative investigation of biological interactions within the NPs (22).

Rawi Ramautar et al. demonstrate a quantitative metabolomics approach to probe the metabolites that bind to NMs, including their co-interaction with proteins from serum to form a complete corona. NMs are promptly coated with biomolecules in biological systems leading to the formation of the so-called corona. To date, research has predominantly focused on the protein corona and how it affects NM uptake, distribution, and bioactivity by conferring a biological identity to NMs enabling interactions with receptors to mediate cellular responses. Thus, protein corona studies are now integral to nanosafety assessment. However, a larger class of molecules, i.e. the metabolites, which are orders of magnitude smaller than proteins (<1,000 Da) and regulate metabolic pathways, has been largely overlooked. This hampers the understanding of bio–nano interface, development of computational predictions of corona formation, and investigations into uptake or toxicity at the cellular level, including identification of molecular initiating events triggering adverse outcome pathways. Here, a capillary electrophoresis–mass spectrometry-based metabolomics approach reveals that pure polar oncogenic metabolite standards differentially adsorb to a range of six NMs (SiO_2, three TiO_2 with different surface chemistries, and naive and carboxylated polystyrene NMs). The metabolite corona composition is quantitatively compared using protein-free and complete plasma samples, revealing that proteins in samples significantly change the composition of the metabolite corona. This key finding provides the basis to include the metabolite corona in future nanosafety endeavors.

Sijie Lin et al. demonstrated the data specimen of nano–bio interaction and exposure conditions in defining the environmental hazard probable of nano-photocatalysts. Nano-photocatalysts are potent for their capability to vitiate pollutants and accomplish water unbearable catalyzed by light. Nanophotocatalysts are representative of TiO_2 and g-C_3N_4 and evaluate the environmental hazard of zebrafish. The short-lived reactive oxygen species generated by nano-photocatalysts only exert injury to the hatched larvae. The input of solar energy determined by the depth of water, irradiation time, and light intensity greatly inspire the toxicity outcome. This thesis interprets the prominence of nano–bio dealings and environmental exposure circumstances in influential safety profile of nano-photocatalysts.

Robert Hurt et al. authenticated the chemical and colloidal dynamics of MnO_2 nanosheets when disseminated in biological media suitable for nanosafety assessment. Many layered crystal phases can be exfoliated or accumulated into ultrathin 2D nanosheets with novel properties not attainable by particulate or fibrous nanoforms. A novel article of MnO_2 is that sensitivity to chemical reduction leads to dissolution of a release of cations. Bio-dissolution is the critical factor in assessing the nanosafety of 2D materials and have a greater prospect on the timing and location of material in the biological/environmental exposures. As MnO_2 is insoluble in aqueous in nature and react with the strong and weak reducing agents in the biological fluid environments. The *in vitro* dissolution is validated and it is slow for cell culture media for MnO_2 internalization by cells in the nanosheets. The bio-dissolution study aids in the prediction of nanotoxicity and nanosafety assessments (23).

The toxicity exhibited due to silver NPs are interpreted by Arno Gutleb et al. They demonstrated the transformation of silver NPs in a food matrix and GI fluids. The physicochemical alterations of size-sorted graphene oxide first transformed the toxicological outcomes in an *in vitro* model of the human intestinal epithelium, which is characterized by Philip Demokritou et al. They represent the synthesis of size-sorted GO of small micrometer size, which represents laser diffraction and field emission scanning electron microscopy that interprets the GO agglomerated in the GI region. X-ray photoelectron spectroscopy reveals that GO makes a covalent bond with N-containing groups on its surface. It is shown that the GO employed in this study undergoes reduction. Toxicological assessments are interpreted by electron microscopy with histological alterations by the increase in reactive oxygen species generation in the sample in contrast to the control (24).

Mikolajczyk et al. explained the correlation between zeta potential and intrinsic physicochemical features of metal oxide NPs through computational study. Nano-QSAR predicts ζ of metal oxide NPs consuming only two descriptors:

1. The spherical size of NPs—a constraint from arithmetical analysis of TEM images
2. The energy of the chief engaged molecular orbital per metal atom—a conjectural descriptor premediated by quantum mechanics at semiempirical level of theory (PM6 method)

The obtained model is then considered by prognostic values and *in silico* estimation of the novel ENP was accomplished. This study is the principal part in sprouting an inclusive and computationally based system to envisage the physicochemical properties which is accountable for the agglomeration and accumulation in the metal oxide NPs (25).

Papa et al. (26) reported linear and nonlinear modeling of the cytotoxicity of TiO$_2$ and ZnO NPs by empirical descriptors. Dissimilar regression techniques were functional and associated with scrutiny of the robustness of the models and their external extrapolative ability. Titanium oxide (TiO$_2$) and zinc oxide (ZnO) NPs proved at characteristic concentration their competence to interject the lipid membrane in cells. This info may be convenient to disguise the prospective for injurious consequence of NP in dissimilar experimental conditions and to enhance the intention of toxicological tests.

11.13 BAYESIAN NETWORK

Interpretation	Method	Type of Nanomaterials	Implementation of Nanotoxicology
A compact, flexible and interpretable representation of a joint probability distribution. It is a useful tool in knowledge discovery as directed acyclic relations between variables	Statistical approach Rigorous method	TiO$_2$, SiO$_2$, Ag, CeO$_2$, and ZnO	Hazard ranking
		Cross-linked iron oxide (CLIO) NPs	Cell-specific targeting
		Small molecules and NPs	Nano-QSAR
		Organic, inorganic, and carbon-based NPs	24 hours post fertilization (hpf) effect on zebrafish model
		ZnO, CuO, Co$_3$O$_4$, and TiO$_2$	Quantitative feature–activity relationships (QFARs)
		SiO$_2$ NPs	Percent cellular viability (CV%) prediction

Interpretation	Method	Type of Nanomaterials	Implementation of Nanotoxicology
It is a toolkit to build products with local AI	Statistical approach	Metal oxide NPs ZnO, CuO, Co$_3$O$_4$, and TiO$_2$ SiO$_2$ NPs	Cell viability assay Quantitative feature–activity relationships (QFARs) Percent cellular viability (CV%) prediction
It is a multichip designed to perform high-speed inferencing for machine learning (ML) models	Rigorous method	All categories of nanomaterial cytotoxicity modeling	External leave-one-out cross-validation (LOO) for methodology authentication

11.14 CONCLUSION

AI has potential applications in the field of biomedicine as they can intercalate the human intelligence and possess enormous applications in the biomedical field. ML is a subset of AI, which accesses data, analyzes trends, and generates intelligent perceptions and has enormous prospective to hasten the expansion and practice of safer NMs for industrial applications. The main apprehensions allotting the field back are still the comparative scantiness of high-quality data which are incorporated to equip, train, and validate models, the necessity for improved mathematical descriptors to encrypt NMs properties, methods, and ways of integrating the heterogeneity and dynamic nature of the "biologically relevant entity" when NMs transit precise biological environments and compartments. Apprehending this complication with vigorous mathematical descriptors is of principal prominence, as descriptor quality is the key element for robust and predictive ML model generation. Progression in automated synthesis and characterization, high-content screening, predictive modeling of the NP corona for a given environment, interpreting of how to mathematically encrypt the bio-physicochemical surface properties of NP, and advances in deep and shallow ML and AI approaches on the horizon should eradicate these roadblocks and hasten a rapid spreading out in the power and practicality of these computational methods in safer design NMs.

ABBREVIATIONS

AI   Artificial Intelligence
ANN   artificial neural network
CNN   convolutional neural network
DL   deep learning
DNN   deep neural network
ENP   engineered nanoparticle
FRET   fluorescence resonance energy transfer
GI   gastrointestinal tract
GO   graphene oxide
HAS   human serum albumin
HUVEC   human umbilical vein endothelial cells
ML   machine learning
NAMs   new approach methods
OECD   Organization for Economic Co-operation and Development
PChem   physiochemical properties
QSAR   quantitative structure–activity relationship
QSPR   quantitative structure–property relationships
RA   risk assessment
REACH   Registration, Evaluation, Authorization and Restriction of Chemicals
SOP   standard operating procedure
TEM   transmission electron microscopy

REFERENCES

1. Zhao, Y., & Nalwa, H. S. (2006) Nanotoxicology: Interactions of Nanomaterials with Biological Systems. California: American Scientific Publishers.
2. Chen, C., Li, Y.-F., Qu, Y., Chai, Z., & Zhao, Y. (2013). Advanced nuclear analytical and related techniques for the growing challenges in nanotoxicology. Chem. Soc. Rev., 42(21), 8266–8303. https://doi.org/10.1039/C3CS60111K
3. Chen, C., Leong, D. T., & Iseult Lynch. Rethinking Nanosafety: Harnessing Progress and Driving Innovation. Small, 16(21), e2002503.
4. Liu, S., & Xia, T. (2020). Continued efforts on nanomaterial environmental health and safety is critical to maintain sustainable growth of nano industry. Small, 16(21), 2000603.
5. Shatkin, J.A. (2020). The future in nanosafety. Nano Lett., 20(3), 1479–1480.
6. Schmid, O., & Stoeger, T. (2016). Surface area is the biologically most effective dose metric for acute nanoparticle toxicity in the lung. J. Aerosol Sci., 99, 133–143.
7. Salieri, B., Turner, D. A., Nowack, B., & Hischier, R. (2018). Life cycle assessment of manufactured nanomaterials: Where are we? NanoImpact, 10, 108–120.
8. Schulte, P. A., Geraci, C. L., Murashov, V., Kuempel, E. D., Zumwalde, R. D., Castranova, V., Hoover, M. D., Hodson, L., & Martinez, K. F. (2014). Occupational safety and health criteria for responsible development of nanotechnology. J. Nanopart. Res., 16, 2153.
9. Savolainen, K., Backman, U., Brouwer, D., Fadeel, B., Fernandes, T., Kuhlbusch, T., Landsiedel, R., & Lynch, L., I. Pylkkänen. (2013). Members of the NanoSafety Cluster, Nanosafety in Europe 2015—2025: Towards Safe and Sustainable Nanomaterials and Nanotechnology Innovations. Helsinki, Finland: Finnish Institute of Occupational Health.
10. Goede, H., Christopher-de Vries, Y., Kuijpers, E., & Fransman, W. (2018). A review of workplace risk management measures for nanomaterials to mitigate inhalation and dermal exposure. Ann. Work Expo. Health, 62(8), 907–922. https://doi.org/10.1093/annweh/wxy032
11. Chen, C., Waller, T., & Walker, S. L. (2017). Visualization of transport and fate of nano and micro-scale particles in porous media: Modeling coupled effects of ionic strength and size. Environ. Sci. Nano, 4(5), 1025–1036. https://doi.org/10.1039/C6EN00558F
12. Copeland, B. J. (2020). Artificial Intelligence. Chicago, IL: Britannica Group, Inc. https://www.britannica.com/technology/artificial-intelligence (accessed January 22, 2021).
13. Lavecchia, A. (2019). Deep learning in drug discovery: Opportunities, challenges and future prospects. Drug Discov Today, 24(10), 2017–2032.
14. Chen, G., Vijver, M. G., Xiao, Y., & Peijnenburg, W. J. G. M. (2017). A review of recent advances towards the development of (quantitative) structure–activity relationships for metallic nanomaterials. Materials, 10, 1013.
15. van Leeuwen, K., Schultz, T. W., Henry, T., Diderich, B., & Veith, G. D. (2009). Using chemical categories to fill data gaps in hazard assessment. SAR QSAR Environ. Res., 20, 207.
16. Puzyn, T., Rasulev, B., Gajewicz, A., Hu, X., Dasari, T. P., Michalkova, A., Hwang, H. M., Toropov, A., Leszczynska, D., & Leszczynski, J. (2011). Using nano-QSAR to predict the cytotoxicity of metal oxide nanoparticles. Nat. Nanotechnol., 6, 175.
17. Szefler, B. (2018). Nanotechnology, from quantum mechanical calculations up to drug delivery. Int. J. Nanomed., 13, 6143.
18. Weissleder, R., Kelly, K., Sun, E. Y., Shtatland, T., & Josephson, L. (2005). Cell-specific targeting of nanoparticles by multivalent attachment of small molecules. Nat. Biotechnol., 23, (11), 1418–1423.
19. Afantitis, A., Melagraki, G., Isigonis, P., Tsoumanis, A., Varsou, D. D., Valsami-Jones, E., Papadiamantis, A., Ellis, L.-J. A., Sarimveis, H., Doganis, P., Karatzas, P., Tsiros, P., Liampa, I., Lobaskin, V., Greco, D., Serra, A., Kinaret, P. A. S., Saarimäki, L. A., Grafström, R., Kohonen, P., Nymark, P., Willighagen, E., Puzyn, T., Rybinska-Fryca, A., Lyubartsev, A., Jensen, K. A. J., Brandenburg, J. G., Lofts, S., Svendsen, C., Harrison, S., Maier, D., Tamm, K., Jänes, J., Sikk, L., Dusinska, M., Longhin, E., Rundén-Pran, E., Mariussen, E., Yamani, N. E., Unger, W., Radnik, J., Tropsha, A., Cohen, Y., Leszczynski, J., Hendren, C. O., Wiesner, M., Winkler, D., Suzuki, N., Yoon, T. H., Choi, J.-S., Sanabria, N., & Gulumian, M., Iseult Lynch. (2020). NanoSolveIT Project: Driving Nanoinformatics research to develop innovative and integrated tools for in silico nanosafety assessment. Comput. Struct. Biotechnol. J., 18, 583–602.
20. Marchese Robinson, R. L., Lynch, I., Peijnenburg, W., Rumble, J., Klaessig, F., Marquardt, C., & Harper, S. L. (2016). How should the completeness and quality of curated nanomaterial data be evaluated? Nanoscale, 8(19), 9919–9943. https://doi.org/10.1039/C5NR08944A

21. Feng, Y., Chang, Y., Xu, K., Zheng, R., Wu, X., & Cheng, Y., & Zhang, H. (2016). Safety-by-design of metal oxide nanoparticles based on the regulation of their energy edges. Small, 16(21), e1907643.

22. Qu, S., Sun, F., Qiao, Z., & Li, J., & Li Shang. (2020). *In situ* investigation on the protein corona formation of quantum dots by using fluorescence resonance energy transfer. Small, 16(21), e1907633.

23. Gray, E. P., Browning, C. L., Vaslet, C. A., Gion, K. D., Green, A., Liu, M., & Kane, A. B., & Hurt, R. H. (2020). Chemical and colloidal dynamics of MnO_2 nanosheets in biological media relevant for nanosafety assessment. Small, 16(21), e2000303.

24. Bitounis, D., Parviz, D., Cao, X., Amadei, C. A., Vecitis, C. D., Sunderland, E. M., Thrall, B. D., Fang, M., & Strano, M. S., & Demokritou, P. (2020). Synthesis and physicochemical transformations of size-sorted graphene oxide during simulated digestion and its toxicological assessment against an *in vitro* model of the human intestinal epithelium. Small, 16(21), e1907640.

25. Mikolajczyk, A., Gajewicz, A., Rasulev, B., Schaeublin, N., Maurer-Gardner, E., Hussain, S., Leszczynski, J., & Puzyn, T. (2015). Zeta potential (ζ) for metal oxide nanoparticles: A predictive model developed by nano-QSPR approach. Chem. Mater., 27(7), 2400–2407.

26. Papa, E., Doucetb, J. P., & Doucet-Panayeb, A. (2015). Linear and non-linear modelling of the cytotoxicity of TiO_2 and ZnO nanoparticles by empirical descriptors. SAR QSAR Environ. Res., 26(7–9), 647–665.

Index

Milton Keynes UK
Ingram Content Group UK Ltd.
UKHW020820141024
449569UK00008B/499